GLOBAL WARMING
IN THE
21ST CENTURY

GLOBAL WARMING IN THE 21ST CENTURY

VOLUME

3 Plants and Animals in Peril

Bruce E. Johansen

Praeger Perspectives

PRAEGER

Westport, Connecticut
London

Library of Congress Cataloging-in-Publication Data

Johansen, Bruce E. (Bruce Elliott), 1950–
    Global warming in the 21st century / Bruce E. Johansen.
        v. cm.
    Includes bibliographical references and index.
    Contents: v.1. Our evolving climate crisis—v.2. Melting ice and warming
seas—v.3. Plants and animals in peril
    ISBN 0-275-98585-7 (set : alk. paper)—ISBN 0-275-98586-5
(v. 1 : alk. paper)—ISBN 0-275-98587-3 (v. 2 : alk. paper)—
ISBN 0-275-99093-1 (v. 3 : alk. paper) 1. Global warming. 2. Global
warming—Environmental aspects. 3. Renewable energy resources.
4. Global warming—Political aspects. I. Title: Global warming in the
twenty-first century. II. Title.
    QC981.8.G56J643   2006
    363.738'74—dc22   2006006633

British Library Cataloguing in Publication Data is available.

Library of Congress Catalog Card Number: 2006006633
ISBN: 0–275–98585–7 (set)
        0–275–98586–5 (vol.1)
        0–275–98587–3 (vol.2)
        0–275–99093–1 (vol.3)

First published in 2006

Praeger Publishers, 88 Post Road West, Westport, CT 06881
An imprint of Greenwood Publishing Group, Inc.
www.praeger.com

Printed in the United States of America

The paper used in this book complies with the
Permanent Paper Standard issued by the National
Information Standards Organization (Z39.48–1984).

10 9 8 7 6 5 4 3 2 1

praying that our grandchildren will be able to endure the world we are shaping for them.

FURTHER READING

Benton, Michael J. *When Life Nearly Died: The Greatest Mass Extinction of All Time.* London: Thames and Hudson, 2003.

Hallam, Anthony, and Paul Wignall. *Mass Extinctions and Their Aftermath.* Oxford: Oxford University Press, 1997.

Schlesinger, James. "The Theology of Global Warming." *Wall Street Journal,* August 8, 2005, A-10.

"Suffocation Suspected for Greatest Mass Extinction." NewScientist.com News Service, September 9, 2003. www.newscientist.com/news/news .jsp?id=ns99994138.

Wolf, Martin. "Hot Air about Global Warming." *London Financial Times,* November 29, 2000, 27.

Given these circumstances, the Kyoto Protocol may be a climatic dead letter, even though its approval by Russia in September 2004 produced worldwide implementation on paper. Russia joined 124 other countries in ratifying the protocol and, with its 17.5 percent share of worldwide carbon dioxide emissions, raised the total signatories to cover slightly more than 60 percent of worldwide emissions, above the 55 percent required to bring Kyoto into force. Seven years after its negotiation in 1997, however, the only sizable countries that have come close to meeting Kyoto Protocol target emission reductions have been Great Britain and Germany. Most other signatories have not met their goals, and many third world countries (India and China among them) are not bound by its provisions.

The Kyoto Protocol has become more of a political rallying cry than a serious challenge to global warming, which is developing much more quickly than any respondent diplomacy. Even if the protocol was to be fully implemented, a projected temperature rise of 2 degrees C by 2050 would be shaved only by 0.07 degree C, according to calculations by atmospheric scientist Thomas M. L. Wigley (Wolf 2000, 27). In other words, the Kyoto goals are only a small fraction of the reduction in emissions required if worldwide temperature levels are to be stabilized during the twenty-first century and afterward.

Governments around the world have dickered over climate-change policy for two decades, while the United States, which produces between one-fourth and one-fifth of the world's greenhouse gases, has ignored the Kyoto Protocol. In the meantime, global emissions of carbon dioxide from fossil fuel combustion increased by 13 percent above 1990 levels by the year 2000, mainly due to huge pollution increases in third world nations, as China and India industrialize rapidly on a power base fueled mainly by cheap, dirty coal. As the atmosphere continues to warm, diplomacy falls further behind. Given these circumstances, humanity is backing itself into a trap, the dimensions of which will not become evident until well after the circumstances of a very miserable future have been set into place. If global warming has any aspects of theology, as James Schlesinger believes, it may be in to warmth. Red squirrels, for example, have adapted their breeding season; insects find new ranges, and bark beetles, whose spread has been notable in Alaska, have been devouring evergreen forests throughout western North America.

Humans, among the most adaptable of animals, cannot fully escape climate change, the subject of Chapter 19. Among the most notable effects are increasing deaths from heat waves in locations lacking the money to adapt with air conditioning. A number of diseases, including malaria and West Nile virus, thrive as warmth expands the ranges of insects that spread them. Air pollution also tends to increase in warmer climates that allow air to stagnate more easily. The burning of greenhouse gases (especially coal) also increases air pollution. Hay fever arrives early and stays late, even as some diseases, such as influenza, fare badly in a warmer world.

These effects may be only an early, pallid indication of the changes awaiting us in the twenty-first century as the pace of warming accelerates to a point where, probably by about the year 2050, it will feed upon itself, altering the nature of weather and the lives of plants, animals, and humans beyond any experience in history or memory. Within a century, coming generations may find themselves wishing for bygone days when many people thought that the comeuppance of the fossil fuel age could be easily contained or ignored.

Chapter 20 concludes with a survey of solutions, from the prosaic (such as changes in building codes), to wind and solar energy, to the possibility that organisms can be invented that will "eat" carbon dioxide. A hard look also is taken at notions of corporate responsibility and the scope of diplomatic measures, such as the Kyoto Protocol.

Global greenhouse gas emissions are rising, and evidence of a warming planet is accumulating much more quickly than world diplomacy has been able to address the situation. The snail-paced nature of consultative diplomacy combines with the reality that we feel the results of fossil fuel effluvia perhaps forty to fifty years after the fact to create a trap in which human responses to global warming take place several decades after nature requires them.

more, substantial warming of a type not seen on Earth for millions of years could provoke mass extinctions of animals with no place left to migrate. Thus, the scientists who are studying previous extinctions are usually not doing so out of pure curiosity but out of concern that humankind may be setting up a similar situation. Doubtless, James Schlesinger sleeps better because he is free of such concerns. Where the climatic rubber meets the ecological road, however, humankind is already pushing many animals toward extinction, with climate change only one of several provocations.

Another set of scientific studies has tested the assumption that elevated levels of carbon dioxide in the atmosphere will benefit plants. Because plants utilize carbon dioxide as part of their growth cycle, some climate "skeptics" have assumed that more of it will be a good thing. The science thus far indicates that this belief is simplistic—and often wrong. In some cases, elevated CO_2 levels actually impede plant growth. In others, they produce more quantity (such as foliage), but less quality (such as seeds). Warming, especially at night, may reduce worldwide yields of rice, a staple crop.

Effects of warming on plants and animals already have become evident around the world. Chapter 17 is devoted to describing records of changes in the British Isles, including earlier first-flowering dates, later freezes, the movement of butterflies, and the spread of scorpions. Chapter 18 is devoted to changes provoked by climate change worldwide in the lives of plants and animals. Many amphibians, for example, have found themselves threatened not only by warming habitats but also by the spreading use of various toxic chemicals and other human-caused insults.

Higher-elevation animals, such as the Rocky Mountain pika, have found themselves without sustaining habitats as warming ascends mountainsides. Some sea birds have starved as warming waters kill their food sources. Every animal has its own ecological adaptations related to temperature. Warmth alters the sex ratio of loggerhead sea turtles, for example, to the point where genders are seriously out of balance, threatening survival of the species. Some animals already have adapted massive amounts of carbon dioxide into the atmosphere, causing a number of feedbacks that accelerated global warming of about 6 degrees C. (Six degrees C is within the forecast range of the IPCC for the end of this century.) In a chapter titled "What Caused the Biggest Catastrophe of All Time?" Benton sketches out how the warming of that time may have fed upon itself:

> The end-Permian runaway greenhouse may have been simple. Release of carbon dioxide from the eruption of the Siberian Traps [volcanoes] led to a rise in global temperatures of 6 degrees C. or so. Cool polar regions became warm and frozen tundra became unfrozen. The melting might have penetrated to the frozen gas hydrate reservoirs located around the polar oceans, and massive volumes of methane may have burst to the surface of the oceans in huge bubbles. This further input of carbon into the atmosphere caused more warming, which could have melted further gas hydrate reservoirs. So the process went on, running faster and faster.
>
> The natural systems that normally reduce carbon dioxide levels could not operate, and eventually the system spiraled out of control, with the biggest crash in the history of life. (pp. 276–277)

Greg Retallack, an expert in ancient soils at the University of Oregon in Eugene, has speculated that this methane belch was of such a magnitude that it could have caused mass extinction of land animals via oxygen starvation. Bob Berner of Yale University has calculated that a cascade of effects on wetlands and coral reefs may have reduced oxygen levels in the atmosphere from 35 percent to just 12 percent in only 20,000 years. Marine life also may have suffocated in the oxygen-poor water ("Suffocation" 2003).

Chapters 16 through 19 of this third volume address the effects of significant global warming on plant and animal life, as well as on human health. Relatively modest warming experienced during the last century (1–2 degrees F in most areas, outside of polar regions) already has driven some animals and plants toward cooler environments. In a century or

INTRODUCTION: VOLUME 3

More Problems and Some Solutions

As this set of books was being readied for production late in 2005, climate change was continuing to engender its usual quota of vehement disagreement and debate. As I read the urgent warnings of many scientists that the world risks a miserably hot future by inaction, I also encountered, in the *Wall Street Journal*, opinions that dismiss the entire idea as "theology" from liberals atoning for their imaginary concepts of "original sin." Who needs solutions, one may ask, for imaginary problems? Ignoring the 35 percent rise in carbon dioxide levels in a century and a half (and its obvious relation to humankind's growing use of fossil fuels), James Schlesinger, who during the 1970s served as the U.S. secretary of energy, wrote that "the burning of fossil fuels . . . [for environmentalists] is the secular counterpart of man's Original Sin" (2005, A-10).

How comforting it must be to assume that our energy paradigm has no relation to an obviously warming atmosphere and that energy industries and consumers bear no blame. Burn oil; be happy.

As I read Schlesinger's polemic, I also encountered the book *When Life Nearly Died: The Greatest Mass Extinction of All Time* (2003), in which Michael J. Benton describes a mass extinction at the end of the Permian period, about 250 million years ago, during which at least 90 percent of life on Earth died. The extinction probably was initiated by massive volcanic eruptions in Siberia. According to present theories, advanced first by Hallam and Wignall (1997), the eruptions injected

ACKNOWLEDGMENTS

Anyone who has written and published a book knows well that it is hardly a solitary journey, even after many hundreds of hours alone at the keyboard. Along the way, many thanks are due, in my case to my wife Pat Keiffer and my family (Shannon, Samantha, Madison), who kept me clothed and fed while enduring numerous bulletins from global warming's many scientific and political fronts. Gratitude also is due to the people of University of Nebraska at Omaha Interlibrary Loan, who can get just about anything that's been published anywhere; to my editors Heather Staines and Lisa Pierce; to the production crew; to the University of Nebraska at Omaha School of Communication Director Jeremy Lipschultz (himself an accomplished author who knows my working habits); and Deans Robert Welk and Shelton Hendricks (for partial relief from teaching duties). Further debts are owed to manuscript reviewers Andrew Lacis of NASA's Goddard Institute for Space Studies in New York City; Gian-Reto Walther of the Institute of Geobotany, University of Hannover, Germany; and Julienne Stroeve, with the Cooperative Institute for Research in Environmental Studies, University of Colorado, Boulder.

Hansen hopes that "the grim 'business-as-usual' climate change" may be avoided by slowing the growth of greenhouse gas emissions during the first quarter of the present century, requiring "strong policy leadership and international cooperation" (Hansen 2005). However, he noted (venturing into the realm of politics) that "special interests have been a roadblock wielding undue influence over policymakers. The special interests seek to maintain short-term profits with little regard to either the long-term impact on the planet that will be inherited by our children and grandchildren or the long-term economic well-being of our country" (Hansen 2005). Hansen leaves to the audience the task of putting names and faces to the special interests who, along with the rest of us, are attending this crucial juncture in the history of the planet and its inhabitants.

FURTHER READING

Hansen, James E. Is There Still Time to Avoid "Dangerous Anthropogenic Interference" with Global Climate? A Tribute to Charles David Keeling. Paper delivered to the American Geophysical Union, San Francisco, December 6, 2005. www.columbia.edu/~jeh1/keeling_talk_and_slides.pdf.
"How to Combat Global Warming: In the End, the Only Real Solution May Be New Energy Technologies," *Business Week*, August 16, 2004, 108.

with fossil fuel CO_2 emissions continuing to increase at about 2 percent a year as in the past decade, yield additional warming of 2 or 3 degrees C this century and imply changes that constitute practically a different planet" (Hansen 2005).

Stop for a moment, and ponder the words, delivered in the measured tones of a veteran scientist: "Practically a different planet," a very real probability by the end of the twenty-first century. Hansen is not joking. He continued: "I present multiple lines of evidence indicating that the Earth's climate is nearing, but has not passed, a tipping point, beyond which it will be impossible to avoid climate change with far-ranging undesirable consequences" (Hansen 2005).

Coming to cases, Hansen described changes that will include:

not only loss of the Arctic as we know it, with all that implies for wildlife and indigenous peoples, but losses on a much vaster scale due to worldwide rising seas. Sea level will increase slowly at first, as losses at the fringes of Greenland and Antarctica due to accelerating ice streams are nearly balanced by increased snowfall and ice sheet thickening in the ice sheet interiors. But as Greenland and West Antarctic ice is softened and lubricated by melt-water and as buttressing ice shelves disappear due to a warming ocean, the balance will tip toward ice loss, thus bringing multiple positive feedbacks into play and causing rapid ice sheet disintegration. The Earth's history suggests that with warming of 2 to 3 degrees C the new equilibrium sea level will include not only most of the ice from Greenland and West Antarctica, but a portion of East Antarctica, raising sea level of the order of 25 meters (80 feet).

To be judicious—we don't want to ruin our case with over-statement—one might allow perhaps two or three centuries for a temperature rise in the atmosphere to express itself as sea-level rise from melting ice. Contrary to lethargic ice sheet models, Hansen suggests, real-world data suggest substantial ice sheet and sea-level change in centuries, not millennia. Now take a look at a map of the world and pay attention to the coastal urban areas. Is anyone worried yet?

comprising the kinds of organic chemicals that are now made from oil and natural gas, the other to generate hydrogen fuel from water and sunshine.

The coming energy revolution will engender economic growth and become an engine of wealth creation for those who realize the opportunities that it offers. Denmark, for example, is making every family a share owner in a burgeoning wind-power industry. The United Kingdom is making plans to reduce its greenhouse-gas emissions 50 percent in 50 years. The British program begins to address the position of the Intergovernmental Panel on Climate Change (IPCC) that emissions will have to fall 60 to 70 percent by century's end to avoid significant warming of the lower atmosphere due to human activities. The Kyoto Protocol, with its reductions of 5 to 15 percent (depending on the country) is barely earnest money compared to the required paradigm change, which will reconstruct the system and the way most of the world's people obtain and use energy.

Solutions will combine scientific achievement and political change. We will end this century with a new energy system, one that acknowledges nature and works with its needs and cycles. Economic development will become congruent with the requirements of sustaining nature. Coming generations will be able to mitigate the effects of greenhouse gases without the increase in poverty so feared by skeptics. Within decades, a new energy paradigm will be enriching us and securing a future that works with the requirements of nature, not against it.

So, how much "wiggle room" does the Earth and its inhabitants have before global warming becomes a truly world-girdling disaster, rather than what many people take to be a set of political, economic, and scientific debating points? Perhaps the best synopsis was provided by James Hansen during a presentation at the American Geophysical Union annual meeting in San Francisco, December 6, 2005. The Earth's temperature, with rapid global warming over the past 30 years, said Hansen, is now passing through the peak level of the Holocene, a period of relatively stable climate that has existed for more than 10,000 years. Further warming of more than 1 degree C "will make the Earth warmer than it has been in a million years. 'Business-as-usual' scenarios, dilemma—solar, wind, hydrogen, and others—will evolve during this century. Within our century, necessity will compel invention. Other technologies may develop that have not, as yet, even broached the realm of present-day science fiction any more than digitized computers had in the days of the Wright Brothers a hundred years ago. We will take this journey because the changing climate, along with our own innate curiosity and creativity, will compel a changing energy paradigm.

Such change will not take place at once. A paradigm change in basic energy technology may require the better part of a century, or longer. Several technologies will evolve together. Oil-based fuels will continue to be used for purposes that require it. (Air transport comes to mind, although engineers already are working on ways to make jet engines more efficient.)

A wide variety of solutions are being pursued around the world, of which the following are only a few examples. Some changes involve localities. Already, several U.S. states are taking actions to limit carbon dioxide emissions despite a lack of support from the U.S. federal government. Building code changes have been enacted. Wind-power incentives have been enacted—even in Bush's home state of Texas, where some oil fields now host wind turbines.

Wind turbines and photovoltaic solar cells are becoming more efficient and competitive. Improvements in farming technology are reducing emissions. Deep-sea sequestration of CO_2 is proceeding in experimental form, but with concerns about this technology's effects on ocean biota. Tokyo, where a powerful urban heat island has intensified the effects of general warming, has proposed a gigantic ocean-water cooling grid. Britain and other countries are considering carbon taxes.

J. Craig Venter, the maverick scientist who compiled a human genetic map with private money, has decided to tap a $100 million research endowment he created from his stock holdings to scour the world's deep ocean trenches for bacteria that might be able to convert carbon dioxide to solid form using little sunlight or other energy. Failing that, Venter proposes to synthesize such organisms via genetic engineering. He would like to invent two synthetic microorganisms: one to consume carbon dioxide and turn it into raw materials

As temperatures rise, energy policy in the United States under the George W. Bush administration generally ignores atmospheric physics. Gas mileage for U.S. internal combustion engines has, in fact, declined during the last two decades, but gains in energy efficiency have been more than offset by increases in vehicle size, notably through sport-utility vehicles. And now comes a mass advertising campaign aimed at security-minded U.S. citizens for the biggest gas-guzzler of all, the fortress-like Humvee. Present-day automotive marketing may seem quaint in a hundred years. By the end of this century, perhaps sooner, the internal combustion engine and the oil (and natural gas) burning furnace will become museum pieces. They will be as antique as the horse and buggy is today. Such change will be beneficial and necessary.

As of 2005, the federal government of the United States (which, as a nation, produces almost one-quarter of the world's greenhouse gases) was sitting out the next worldwide energy revolution. The United States is being led (if that is the word) by a group of minds still set to the clock of the early twentieth-century fossil fuel boom. The Bush administration not only has refused to endorse the Kyoto Protocol, but also has (with a few exceptions, such as its endorsement of hydrogen and hybrid-fueled automobiles) failed to take seriously the coming revolution in the technology of energy production and use. In a century, George Bush's bust may sit in a greenhouse-gas museum, not far from a model of an antique internal-combustion engine. A plaque may mention his family's intimate ties to the oil industry as a factor in his refusal to think outside that particular box.

As the White House banters about "sound science," yellow-jacket wasps were sighted on Northern Baffin Island during the summer of 2004. By the end of the twenty-first century, if "business as usual" fossil fuel consumption is not curbed substantially, the atmosphere's carbon dioxide level will reach 800 to 1,000 parts per million. The last time this level was reached 55 million years ago, during the days of the dinosaurs, the water at the North Pole, then devoid of ice, reached approximately 68 degrees F.

Before the end of this century, the urgency of global warming will become manifest to everyone. Solutions to our fossil fuel sources, along with the accelerating climate change from greenhouse gases accumulating in the atmosphere. According to an editorial in *Business Week*, "A national policy that cuts fossil-fuel consumption converges with a geopolitical policy of reducing energy dependence on Middle East oil. Reducing carbon dioxide emissions is no longer just a 'green' thing. It makes business and foreign policy sense, as well. . . . In the end, the only real solution may be new energy technologies. There has been little innovation in energy since the internal combustion engine was invented in the 1860s and Thomas Edison built his first commercial electric generating plant in 1882" ("How to Combat Global Warming" 2004, 108).

Even the climate skeptics don't deny that climate has warmed. Temperatures have, indeed, been rising. Nine of the ten warmest years worldwide have been recorded since 1990. General warming does not imply that all cold weather has ended—instead, extremes have generally been increasing. As global averages have increased, for example, a spell of intense cold killed more than 1,000 people in India and Bangladesh during the winter of 2002–2003, only a few months after hundreds perished of record heat in the same area.

Increasing evidence also indicates that rising temperatures are changing the hydrological cycle, helping to cause intensifying chances of precipitation extremes of drought and deluge. Western Europe has experienced flooding rains while the western interior of North America has been suffering what may be the worst drought since that of a thousand years ago, which ruined the civilization of the Anasazis. Intensity of storms often increases with warmth. In the midst of drought during the summer of 2002, for example, sections of Nebraska experienced cloudbursts that eroded soil and washed out an interstate highway. Hours later, the drought returned.

Temperature also does not fully express itself immediately, but through a feedback loop of perhaps a half-century. We are, thus, now experiencing the climate related to greenhouse gas levels of about 1960. Since that time, carbon dioxide levels have risen substantially, all but guaranteeing further, substantial warming during at least the next half-century.

PREFACE

The Next Energy Revolution

In one hundred years, students of history may remark at the nature of the fears that stalled responses to climate change early in the twenty-first century. Skeptics of global warming kept change at bay, it may be noted, by appealing to most people's fear of change that might erode their comfort and employment security, all of which were psychologically wedded to the massive burning of fossil fuels. A necessary change in our energy base may have been stalled, they might conclude, beyond the point where climate change forced attention, comprehension, and action.

Technological change always generates fear of unemployment. Paradoxically, such changes also generate economic activity. A change in our basic energy paradigm during the twenty-first century will not cause the ruination of our economic base, as some skeptics of climate change believe, any more than the coming of the railroads in the nineteenth century ruined an economy in which the horse was the major land-based vehicle of transportation. The advent of mass automobile ownership early in the twentieth century propelled economic growth, as did the transformation of information gathering and handling via computers in the recent past. The same developments also put blacksmiths, keepers of hand-drawn accounting ledgers, and anyone who repaired manual typesetters out of work.

We are overdue for an energy system paradigm shift. Limited oil supply and its location in the volatile Middle East make a case for new

Color insert in Volume 2 precedes Part IV.

Contents

CONTENTS

887015

V FLORA AND FAUNA

INTRODUCTION

Global warming is one among several human provocations that have been contributing to large-scale alterations in the flora and fauna that the anthropomorphic economy places outside its ambit—that is, "wild" populations. The growth of human populations around the world, along with attendant pollution and loss of habitat, has, during the twentieth and twenty-first centuries, set the stage for one of a handful of mass extinction episodes in the geophysical history of the Earth.

"The biotic response to 30 years of enhanced global warming [1970–2000] has become perceptible and substantial," writes Gian-Reto Walther, who has published several scientific "meta studies" that evaluate hundreds of specific articles addressing the response of flora and fauna to changing climatic conditions (2003, 177). Walther's surveys range the world, describing increasing stands of broadleaf evergreen trees in Switzerland (2002, 129–139) and the migration of frost-sensitive tropical plants up mountains in Hong Kong, as well as the poleward shift of holly in Scandinavia (2003, 173), to cite three examples of many.

Scientists have examined periods in the distant past in which spikes in worldwide temperatures from natural causes led to widespread extinctions of flora and fauna. These episodes are not being studied as academic exercises but as cautionary lessons in what can happen when Earth's climate heats suddenly, as is expected via global warming during the twenty-first century. Along this road, other scientists have been exploring how "enhanced" (that is, unusually elevated) levels of carbon dioxide and other greenhouse gases affect plant growth and reproduction. Effects on the behavior and reproduction of many animal species, including human beings, already were becoming evident at the beginning of the century. What follows is a survey of a world in flux, with the major changes in the lives of Earth's plants and animals still to come.

16 MASS EXTINCTIONS AND OTHER CONSIDERATIONS

INTRODUCTION: MASS EXTINCTIONS WITHIN A CENTURY?

We are in the midst of one of the Earth's most intense, rapid, and pervasive mass extinctions, which has placed in harm's way many flora and fauna that humankind does not eat or keep as pets. The Earth has experienced mass extinctions before, but all of them have resulted from natural causes. Global warming is one product of humankind's increasing dominance of the Earth that is devastating the native habitats of many animals and plants, driving many to extinction. Compared to past mass extinctions, which were driven by natural catastrophes such as meteor strikes or large-scale volcanism, the present-day human-driven wave of extinctions has been occurring with frightening speed. Given projected rises in temperature during decades to come, the flora and fauna of our home planet thus far have seen only their initial travails.

In the first study of its kind, researchers in a range of habitats including northern Britain, the wet tropics of northeastern Australia, and the Mexican desert said early in 2004 that, given "mid-range" climate change scenarios for 2050, they anticipate that 15 to 37 percent of the species in their sample of regions (covering 20 percent of Earth's surface) will be "committed to extinction" (Thomas et al. 2004, 145). The severity of extinctions is expected to vary with the degree of warming. The study used U.N. projections that world average temperatures will rise 2.5 to 10.4 degrees F by 2100. "We're not talking about the occasional extinction—we're talking about 1.25 million species. It's a massive number," the authors of this study wrote (Gugliotta 2004, A-1).

This study, described in *Nature*, marked the first time scientists have produced a global analysis with concrete estimates of the effect of climate change on many various animal and plant habitats. Thomas led a nineteen-member international team that surveyed habitat decline for 1,103 plant and animal species in Europe; Queensland, Australia; Mexico's Chihuahua Desert; the Brazilian Amazon; and the Cape Floristic Region at South Africa's southern tip (Gugliotta 2004, A-1).

According to the researchers, climate change during the past thirty years already has produced many shifts in the distribution and abundance of plants and animals. Climate change thus has become a major driver of biodiversity change. The survey team used one of ecology's few iron-clad laws: the species-area relationship, first postulated by Charles Darwin in his *Origin of Species* (1859), which holds that a smaller habitable area will host a smaller number of viable species. They then projected habitat changes based on various warming scenarios. With a temperature rise of 0.8 to 1.7 degrees C, they anticipated an 18 percent extinction rate; at 1.8 to 2.0 degrees C, they projected 24 percent, and at more than 2.0 degrees C, 35 percent (Pounds and Puschendorf 2004, 108).

This study emphasizes that examining possible extinctions solely in light of global warming probably understates their potential because, in the real world, they also may be caused by such factors as landscape modification, species invasions, and pollution. By projecting effects of rising temperatures alone, the researchers also realized that they may be ignoring effects of changes in precipitation on habitat. Declines of amphibians in Costa Rica have been traced to reduced cloudiness in mountainous areas that probably are related to warming, for example.

The authors of this study considered a range of possibilities based on the ability of each species to move to a more congenial habitat to escape warming. If all species were able to move, or "disperse," the study said, only 15 percent would be irrevocably headed for extinction by 2050. If no species were able to move, the extinction rate could rise as high as

37 percent (Gugliotta 2004, A-1). The scientists concluded, "These estimates show the importance of rapid implementation of technologies to decrease greenhouse-gas emissions and strategies for carbon sequestration" (Thomas et al. 2004, 145).

The survey's findings were disputed by skeptics, such as William O'Keefe, president of the George C. Marshall Institute, who said that the research "ignored species' ability to adapt to higher temperatures" and assumed that technologies will not arise to reduce emissions (Gugliotta 2004, A-1). Some animals and plants are adapting to warming, to a point, as they move upward in elevation or toward the poles in direction, until something blocks their way. In the European Alps, for example, some plant species have been migrating upward by one to four meters each decade (Grabherr, Gottfried, and Pauli 1994, 448). Across Europe, the growing season in controlled mixed-species gardens lengthened by 10.8 days between 1959 and 1993 (Menzel and Fabian, 1999, 659). In Europe and North America, many migratory birds now arrive earlier in the spring and depart later in the autumn. Butterflies, beetles, dragonflies, and other species are now found further north, where conditions previously were too cold for their survival.

CLIMATE CHANGE AND GLOBAL BIODIVERSITY

Studies by wildlife advocacy groups support many scientists' projections of pending mass extinctions. These studies usually define the problem in terms of declining biodiversity. The studies do more than document present and potential extinctions of several species; they also advocate emphatic action to combat climate change. These groups call on U.S. lawmakers to help curtail greenhouse gas emissions, for example, by enacting higher fuel-efficiency standards for ground transportation and more energy-efficient building codes (Lazaroff 2001).

Two such reports were compiled by the National Wildlife Federation in the United States and the International World Wildlife Fund (WWF), both of which contend that species from the tropics to the poles are at risk. Many species may be unable to move to new areas quickly enough to survive changes that rising temperatures will bring to their historic habitats. Based on a doubling of atmospheric carbon dioxide levels, expected by many scientists within a century, the WWF report (www.worldwildlife.org/news) asserted that one-fifth of the world's most vulnerable natural areas may be facing a "catastrophic" loss of species (Lazaroff 2001). "It is shocking to see that many of our most biologically valuable ecosystems are at special risk from global warming. If we don't do something to reverse this frightening trend, it would mean extinction for thousands of species," said Jay Malcolm, author of the WWF report and a professor at the University of Toronto (Lazaroff 2001).

According to this WWF report, areas most vulnerable to devastation from global warming include species in the Canadian low Arctic tundra; the central Andean dry puna of Chile, Argentina, and Bolivia; the Ural Mountains; the Daurian Steppe of Mongolia and Russia; the Terai-Duar savannah of northeastern India; southwestern Australia; and the fynbos of South Africa. Among the U.S. ecosystems at risk, areas in California, the Pacific Northwest and the northern prairie may be hardest hit, the WWF said. The changes could devastate the shrub and woodland areas that stretch from southern California to San Francisco, prairies in the northern heart of the United States, the Sierra Nevada, the Klamath-Siskiyou forest near the California-Oregon border, and the Sonoran-Baja deserts across the southwestern United States.

Even during the twentieth century, warmer weather (especially milder winters) has been associated with changes in forest species. In one study, Gian-Reto Walther documented the increasing growth of broad-leaved evergreen trees in lower areas of southern Switzerland, where the growing season had increased to as long as eleven months in some areas by roughly the year 2000 (Walther 2002, 129). These areas have remarkably precise records of frost-free days. Between 1902 and 1998, for example, the onset of winter (the first day of frost) at Lugano had changed from an average of the second week in November to the first week of December. The last day of frost had changed from an average of the fourth week in March to the fourth week of February (p. 132).

Research compiled by the National Wildlife Federation (NWF) suggests that global warming will probably pose an intensifying threat to U.S. wildlife, including more trouble with invasive species, and that significant environmental changes will jeopardize human quality of life in the near future (Lazaroff 2001). "Global warming has come down to Earth for the wildlife right in our backyards," said Mark Van Putten, president of NWF. "The effects are already happening and will likely worsen unless we get serious about reducing emissions of carbon dioxide and other heat-trapping gases to help slow global warming" (Lazaroff 2001). The NWF's findings appeared in a book, *Wildlife Responses to Climate Change* (2001), edited by Stephen Schneider of Stanford University and Terry Root of the University of Michigan, that includes eight case studies by researchers that demonstrate how global warming and associated climate change is affecting North American wildlife.

The NWF report asserted that invasive species such as tamarisk shrubs in the U.S. Southwest may expand their ranges, reducing water and food available to native wildlife and humans. Imported fire ants in the Southeast also may expand their range, dominating native ant species and creating an enhanced health risk to humans (Lazaroff 2001). Species such as the sachem skipper butterfly in the Pacific Northwest and the Bay checkerspot butterfly in California already are responding to climatic and weather changes. Changes in climate also may alter habitat for grizzly bears, red squirrels, and other wildlife in the Greater Yellowstone ecosystem region of Wyoming, Montana, and Idaho by contributing to a reduction in whitebark pine trees, an important food source for animals.

Roughly 70 percent of the plants and animals on the fynbos of southern Africa are unique to that area, according to the WWF. In summer, the area is often so parched by drought and heat that it is ravaged by fire, with many plants dependent on regular fires to stimulate seeding. Almost half of this area will become uninhabitable for its major species within a century, according to the WWF (Browne 2002, 15). Some species, such as the springbok of southern Africa, are very attached to their habitats and are expected to die rather than move.

"Migration routes have become increasingly impassable due to human activities," wrote Gian-Reto Walther, summarizing several scientific studies (2003, 177). At the same time, human transportation facilitates invasive species worldwide. Climate change also plays favorites among flora and fauna by removing climatic constraints; witness the explosion of spruce budworms throughout the pine forests of North America (see Chapter 10 in this volume).

"Many habitats will change at a rate approximately 10 times faster than the most rapid changes since the last ice age, causing extinctions," scientist Collier said (Browne 2002, 15). Washington State's Olympic Peninsula rainforest also is on the WWF's list of endangered habitats. The rainforest cannot move as the climate changes. The peninsula is likely to undergo a drastic change, the report said, although scientists say it's difficult to determine how quickly that change will occur.

The Wildlife Society, an association of nearly 9,000 wildlife managers, research scientists, biologists, and educators based in Washington, D.C., forecasts in its report, *Global Climate Change and Wildlife in North America*, that many animals may find their migratory paths blocked by cities, transportation corridors, or farmland. Predators and their prey may not be able to move at the same time, impeding natural balances. Plants could suffer if the birds and insects that pollinate them head to cooler climes. Wetlands in the Midwest and central Canada are expected to dry up, causing some duck species to decline by as much as 69 percent over the next seventy-five years. Nesting habitat could be lost as wetlands become more suitable for row crops.

Even now, ring-tailed possums are falling dead out of trees in Australia's far north Queensland because the climate has become too hot for them. Green ring-tailed possums can't survive more than five hours at an air temperature above 30 degrees C (O'Malley 2003, 14). The ring-tailed possums are not alone. Australia's Rainforest Cooperative Research Centre anticipates that half of all the unique mammals, reptiles, and birds in far north Queensland's rainforests could become extinct given a 3.5 degree C rise in temperature, which is the midpoint of the Intergovernmental Panel on Climate Change's (IPCC's) range of estimated temperature increase for the end of the twenty-first century. Most of these animals are found only in the Wet Tropics World Heritage Area near Cairns, at elevations above 600 meters (p. 14).

GLOBAL WARMING AND MASS EXTINCTIONS: WHAT HAPPENED 250 MILLION YEARS AGO?

As the anticipation of mass extinctions abetted by global warming has become more common, scientists have sought paleoclimatic parallels. According to some scientists, the worst mass extinction in the history of the planet could be replicated in as little as a century if global warming continues at the pace forecast by the Intergovernmental Panel on Climate Change (IPCC). Researchers at England's Bristol University have estimated that a 6 degree C increase in global temperatures was enough to play a role in the annihilation of up to 95 percent of the species alive on Earth at the end of the Permian period 251 million years ago, roughly the same amount of warming expected by the IPCC if levels of greenhouse gases in the atmosphere continue to rise at present rates (Reynolds 2003, 6).

The wave of mass extinctions at the end of the Permian period probably was caused by a series of very large volcanic eruptions that triggered a runaway greenhouse effect that nearly extinguished life on Earth. Conditions in what geologists have termed this "post-apocalyptic greenhouse" were so severe that 100 million years passed before species diversity returned to former levels. Michael Benton, head of Earth sciences at Bristol University, commented: "The end-Permian crisis nearly marked the end of life. It's estimated that fewer than one in ten species survived. Geologists are only now coming to appreciate the severity of this global catastrophe and to understand how and why so many species died out so quickly" (Reynolds 2003, 6).

The Permian heat wave was felt first and most intensely in the tropical latitudes; loss of species diversity spread from there. Reduction of vegetation, soil erosion, and the effects of increasing rainfall wiped out the lush, diverse habitats of the tropics, which would today lead to the loss of animals such as hippos, elephants, and all of the primates, according to Benton (Reynolds 2003, 6). He added, "The end-Permian extinction event is a good model for what might happen in the future because it was fairly non-specific. The sequence of what happened then is different from today because then the carbon dioxide came from massive volcanic eruptions, whereas today it is coming from industrial activity. However, it doesn't matter where this gas comes from; the fact is that if it is pumped into the atmosphere in high volumes, then that gives us the greenhouse effect and leads to the warming with all the other consequences" (p. 6).

According to present theories, first advanced by Anthony Hallam and Paul Wignall (1997), the volcanic eruptions 251 million years ago provoked a number of biotic feedbacks that accelerated a global warming of about 6 degrees C. In a chapter of his book *When Life Nearly Died: The Greatest Mass Extinction of All Time* (2003) titled "What Caused the Biggest Catastrophe of All Time?" Benton sketched out how the warming (which was accompanied by anoxia) may have fed upon itself:

> The end-Permian runaway greenhouse may have been simple. Release of carbon dioxide from the eruption of the Siberian Traps [volcanoes] led to a rise in global temperatures of 6 degrees C. or so. Cool polar regions became warm and frozen tundra became unfrozen. The melting might have penetrated to the frozen gas hydrate reservoirs located around the polar oceans, and massive volumes of methane may have burst to the surface of the oceans in huge bubbles. This further input of carbon into the atmosphere caused more warming, which could have melted further gas hydrate reservoirs. So the process went on, running faster and faster.
>
> The natural systems that normally reduce carbon dioxide levels could not operate, and eventually the system spiraled out of control, with the biggest crash in the history of life. (pp. 276–277)

Greg Retallack, an expert in ancient soils at the University of Oregon in Eugene, has speculated that this same methane "belch" was of such a magnitude that it caused mass extinction via oxygen starvation of land animals. Bob Berner of Yale University has calculated that a cascade of effects on wetlands and coral reefs may have reduced oxygen levels in the atmosphere from 35 percent to just 12 percent in only 20,000 years. Marine life also may have suffocated in the oxygen-poor water ("Suffocation" 2003). One animal, the meter-long reptile *Lystrosaurus*, survived because it had evolved to live in burrows, where oxygen levels are low and carbon dioxide levels high. According to a report by the New Scientist News Service, "It had developed a barrel chest, thick ribs, enlarged lungs, a muscular diaphragm and short internal nostrils to get the oxygen it needed ("Suffocation" 2003).

According to Chris Lavers, writing in *Why Elephants Have Big Ears* (2000), a spike in worldwide warming contributed to this mass extinction, in part because all of the Earth's continents at the time were combined into one land mass (p. 231). Warming of tropical regions during this period has been estimated at about 11 degrees F, with larger rises near the poles, resulting in a generally warm atmosphere planetwide—"a flattening of the temperature difference between the poles and the equator" (p. 232). This condition, Lavers suspects, drastically slowed or shut down ocean mixing (see Volume 2, Part IV, "Warming Seas," on thermohaline circulation), killing many sea creatures. "Unstirred," wrote Lavers, "the oceans begin to stagnate. Deep waters gradually lost oxygen, and species began to vanish" (p. 233).

What caused this spike in temperatures? The prime suspects, at least in the beginning, are the coal-bearing deposits in the southern reaches of Pangea (the Earth's single land mass), which were oxidized after they were lifted by tectonic activity, releasing large volumes of carbon dioxide when the volcanoes erupted. The level of greenhouse gases in the atmosphere thus increased due to the most concentrated bout of volcanic activity on Earth during the last 600 million years. "This injection of volcanic CO_2," wrote Lavers, "was probably the decisive event that ultimately tipped the biosphere into the new era of the Mesozoic" (p. 235).

Gregory Ryskin, a geologist at Northwestern University, has asserted that a smaller-scale methane "burp" may help explain the biblical flood navigated by Noah's Ark. The biblical flood, according to Ryskin

(2003), may be attributable to a methane "burp" from Europe's stagnant Black Sea. Some geological evidence suggests such an event of this type 7,000 to 8,000 years ago. According to an account by Tom Clarke carried in *Nature* (2003), "Ryskin contends that methane from bacterial decay or from frozen methane hydrates in deep oceans began to be released. Under the enormous pressure from water above, the gas dissolved in the water at the bottom of the ocean and was trapped there as its concentration grew." A single disturbance, according to this account, "a small meteorite impact or even a fast-moving mammal, could then have brought the gas-saturated water closer to the surface. Here it would have bubbled out of solution under the reduced pressure. Thereafter the process would have been unstoppable: a huge overturning of the water layers would have released a vast belch of methane."

About 80 percent of all Earth's species died at the end of the Triassic Period and the beginning of the Jurassic. These extinctions have been attributed to the eruption of flood basalts that are said to have released substantial amounts of carbon dioxide into the atmosphere, leading to a swift and catastrophic greenhouse warming. Estimates of carbon dioxide levels at that time have ranged as high as 2,000 to 4,000 parts per million. Models utilized by Lawrence H. Tanner and colleagues suggest that this high level of carbon dioxide was not the case, however. These scholars reported that atmospheric carbon dioxide levels probably did not exceed 250 ppm during this period, roughly the level in recent times before the industrial age. Tanner and colleagues commented: "The relative stability of atmospheric CO_2 across this boundary suggests that environmental degradation and extinctions during the early Jurassic were not caused by volcanic outgassing of CO_2. Other volcanic effects—such as the release of atmospheric aerosols or techtonically driven sea-level change—may have been responsible for this event" (2001, 675).

ALTERATIONS IN FLORA AND FAUNA WITH SMALL TEMPERATURE CHANGES

Even with a global temperature rise of only 0.6 degrees C during the twentieth century, a small fraction of the warming expected in years to come, significant alterations in flora and fauna have been noticed around the Earth, according to one of the most detailed ecological studies of climate change. An international team of scientists working across a range of disciplines found "a major imprint on wildlife" during the twentieth century (Connor 2002, 11).

Writing in *Nature*, Gian-Reto Walther and colleagues said:

There is now ample evidence of the ecological impacts of recent climate change, from polar terrestrial to tropical marine environments. The response of both flora and fauna span an array of ecosystems and organizational hierarchies, from the species to the community levels. Despite continued uncertainty as to community and ecosystem trajectories under global change, our review exposes a coherent pattern of ecological change across systems. Although we are only at an early stage in the projected trends of global warming, ecological responses to recent climate change are already clearly visible. (Walther et al. 2002, 389)

Walther, an ecologist at the University of Hanover in Germany and lead author of the study, said that "we want to emphasize that climate-change impacts are not something we expect for the future but something that is already happening. We are convinced of that. If you have so many studies from so many regions with so many different species involved and all pointing to the same direction of warmer temperatures, then to me it's quite convincing" (Connor 2002, 11). The study included work by British specialists in amphibian breeding cycles and Antarctic ecology, German experts on bird migration, and Australian marine biologists studying changes to coral reefs in tropical oceans.

Walther continued, "It's the first time that researchers from various disciplines have come together to compare their own work. We made comparisons between and across various ecosystems and we compared different species to look for common traits—and they all point to the same direction of warmer temperatures. We know that the global average temperature has increased by 0.6 degrees C. For many people this may sound very minor, but we are quite surprised that this minor change has had so many impacts already on natural ecosystems" (Connor 2002, 11).

All of the scientists found that typical springtime activities, such as the arrival and breeding of migrant birds or the first appearance of butterflies and plants, have occurred progressively earlier during the past forty years. Meanwhile, some warm-weather species and diseases have extended their ranges. "There is much evidence that a steady rise in annual temperatures has been associated with expanding mosquito-borne disease in the highlands of Asia, East Africa and Latin America," the scientists said (Connor 2002, 11).

To cite a few examples among many: mosses and other plants have begun to grow in parts of the Antarctic that were previously considered too cold for such life. Many coral reefs around the world have undergone mass "bleaching" on at least six occasions since 1979, all of which is related to warmer sea temperatures (see Volume 2, Part IV, "Warming Seas"). Changes in wind patterns over the Bering Sea and their interaction with local patterns of ocean circulation have affected the distribution of walleye pollock, an important "forage species" for other fish and sea mammals (Connor 2002, 11). Similar changes in ocean circulation around the Antarctic Peninsula have influenced the breeding range of the krill, an important shrimp-like animal that constitutes the base of the Antarctic food chain (see Volume 2, Part III, "Icemelt around the World").

In Britain, warmer winters have caused newts to breed earlier, bringing them into contact with the eggs and young of the common frog at a point when they are most vulnerable to predation, said Trevor Beebee, professor of molecular biology at the University of Sussex and a coauthor of the Walther study. "Newts and other amphibians with a protracted breeding season are responding to climate change whereas frogs and toads which usually breed earlier are not," Beebee said. "You can relate these changes to temperature changes observed over the same period" (Connor 2002, 11).

Common changes observed by this team of scientists included earlier spring breeding and first singing of birds, earlier arrival of migrating birds, earlier first appearance of butterflies, earlier spawning in amphibians, as well as earlier shooting and flowering of plants. In general, according to this report, spring activities have occurred progressively earlier since the 1960s, at a rate that is easily observable within a single human lifespan (Walther et al. 2002, 389).

SPECIES MOVING TOWARD THE POLES

Other scientific studies indicate that species are generally moving toward the poles—northward in the northern hemisphere and (with a smaller number of examples) southward below the equator. Two studies involving several thousand plant and animal species throughout the world, from plankton to polar bears, provide ample evidence that climate change is reshaping animal and plant habitats at an accelerating rate.

Both studies suggest that habitat change is already well under way. "There is a very strong signal from across all regions of the world that the globe is warming," said Terry Root of Stanford University's Center for Environmental Science and Policy, who headed one of the research teams. "Thermometers can tell us that the Earth is warming, but the plants and animals are telling us that global warming is already having a discernible impact. People who don't believe it should take their heads out of the sand and look around" (Toner 2003, 1-A).

Root and colleagues, in their "meta-analysis" of 143 studies, concluded:

More than 80 per cent of the species that show changes are shifting in the direction expected on the basis of known physiological constraints of species. Consequently, the balance of evidence from these studies strongly suggests that a significant impact of global warming is already discernable in animal and plant populations. The synergism of rapid temperature rise and other stresses, in particular habitat destruction, could easily disrupt the connectedness among species and lead to a reformulation of species communities, reflecting differential changes in species, and to numerous extirpations and possibly extinctions. (Root et al. 2003, 57)

Root's team, and another headed by Camille Parmesan, a biologist at the University of Texas at Austin, looked at hundreds of existing studies of more than 2,000 plants and animals—from shrimp, crabs, and barnacles off the Pacific coast of California to cardinals that nest in Wisconsin—to see whether they could spot the "fingerprint" of changing global temperatures (Toner 2003, 1-A). Parmesan and Gary Yohe calculated that range shifts toward the poles had averaged 6.1 kilometers per decade, with an advance in the beginning of spring by 2.3 days per decade (Parmesan and Yohe 2003, 37).

According to Parmesan and Yohe's analysis, some birds and butterflies had shifted as many as 600 miles northward. Grasses, trees, and other species that lack mobility have moved shorter distances. In a finding that resembled other studies, they found that springtime behavior was occurring earlier, in this case by more than two days per decade. A study of more than 21,000 swallow nests in North America, for example, indicated that the species was laying its eggs nine days earlier in the year 2000 than in 1960. In mountains, various species responded to warming by moving up in elevation, seeking habitats that were similar to areas they formerly had occupied at lower levels. In some areas, such as the Great Smoky Mountains, the mountain peaks had become shrinking "islands" of cool-weather species. The Root and Parmesan studies both indicated that the largest changes in temperature as well as in habitat movement were taking place in the polar regions, where temperature increases have been greatest (see Volume 2, Part III, "Icemelt around the World"). "The most remarkable thing is that we have seen so many changes in so many parts of the world from a relatively small increase in temperature," Root told Mike Toner of the *Atlanta Journal-Constitution*. "When you consider that some people are predicting warming that would be 10 times greater by the end of this century, it's spooky to think about what the consequences might be" (Toner 2003, 1-A).

Examples of seasonal and geographical habitat change due to warming temperatures abound, according to the news report describing

Parmesan and Root's work. For example, the North American tree swallow is among those bird species beginning springtime activities earlier than historically recorded. Field biologists who kept track of roughly 21,000 tree swallow nests in the United States and Canada over forty years concluded that the average egg-laying date has advanced by nine days. Studies in Colorado found that marmots are ending their hibernation about three weeks sooner than during the late 1970s. Measurements taken in Alaska also revealed that growth of white spruce trees has been stunted in recent years, another expected consequence of a rapidly warming climate (Toner 2003, 1-A).

ELEVATED CARBON DIOXIDE LEVELS MAY IMPEDE PLANT GROWTH

Because plants require carbon dioxide to grow, folk wisdom holds that increasing the levels of this trace gas in the atmosphere will accelerate plant growth. Climate change skeptics who contend that higher carbon dioxide levels will benefit plants often ignore the possibility that heat stress or drought-and-deluge conditions, both probable side effects of warming, will cause problems. In addition, scientific support for the idea that elevated carbon dioxide levels aid plant growth is far from unanimous. Although a few studies indicate that in the short term higher carbon dioxide levels and warmer temperatures may help "green" the Earth, longer-run studies cast considerable doubt on the long-term results.

In the short run, rising temperatures and higher levels of carbon dioxide in the atmosphere are increasing vegetative cover in some areas, according to one study (Pegg 2003). Global changes in temperature, rainfall, and cloud cover have given plants more heat, water, and sunlight in areas where climatic conditions once limited growth, according to the study, which was jointly funded by the U.S. National Aeronautics and Space Administration (NASA) and the U.S. Department of Energy. According to a report by the Environment News Service, "The study finds that in general, for the period 1982 to 1999, [in] areas where temperatures restricted plant growth, it became warmer; where sunlight was needed, clouds dissipated; and where it was too dry, more rain fell." Lead author Ramakrishna Nemani, a professor in the forestry school at the University of Montana, said that the study indicates climatic change is "the leading cause for the increases in plant growth over the last two decades." Nemani and colleagues analyzed satellite data and determined that warmer temperatures as well as shifting precipitation patterns and cloud cover led to a 6 percent increase in the amount of carbon stored in plants worldwide. Growth in the Amazon rainforests accounted for nearly half the global increase found in the study, growth linked to reduced cloud cover and steady rainfall.

After three years, according to one set of tests, elevated carbon dioxide and nitrogen deposition together reduced plant diversity, whereas elevated precipitation alone increased it and warmer temperatures had no significant effect. According to this study, "Results show that climate and atmospheric changes can rapidly alter biological diversity, with combined effects that, at least in some settings, are simple, additive combinations of single-factor effects" (Zavaleta et al. 2003, 7650). This research found that doubling the level of carbon dioxide reduced numbers of wildflowers by 20 percent and cut overall plant diversity by 8 percent (Connor 2003). Results from experiments involving potted plants have limited utility, however, and often cannot be extrapolated to the natural environment.

Scientists from Stanford University manipulated levels of carbon dioxide for thirty-six open-air plots of land on which wild plants grew for three years. They doubled carbon dioxide, increased available moisture by 50 percent, caused average temperatures to rise by 1.7 degrees C, and added nitrogen to the soil. All of these conditions could become standard fare in the open air within a century, according to projections of the IPCC. Plots that received all four treatments suffered a decline of 25 percent in volume of wildflowers; those given extra nitrogen or carbon dioxide suffered a 10 or 20 percent decline. Only increased watering produced a rise in diversity. The scientists suggested that increased carbon dioxide, temperature, and nitrogen allowed some plants to grow faster for longer periods but impeded growth (and sometimes survival) of other plants (Connor 2003).

Christopher Field mentioned that the team was surprised to find that increased carbon dioxide and watering caused such disparate effects, given that they were both essential for plants' growth. "One hypothesis is that elevated carbon dioxide added moisture to the soil, which tended to extend the growing season of the dominant plants, leaving less room for other species to grow," Professor Field said. Field's colleague, Erika Zavaleta, said that the study demonstrated how some wild plants would suffer whereas other may benefit from the effects of climate change. "Certain kinds of species are much more sensitive to climate and atmospheric changes than others. It turned out that wildflowers were much more sensitive to the treatments than grasses were, no matter what combination of treatments we tried," Zavaleta described (Connor 2003).

When enhanced temperature, nitrogen, and water were applied to a plot, the production soared by 84 percent, Shaw, another author of the Stanford study, stated. But when carbon dioxide was added to this mix, the production dropped by 40 percent. "This was unexpected," Shaw said. "We think that by applying all four elements in combination in a realistic situation, some other nutrient becomes a limiting factor to growth" (Recar 2002; Shaw et al. 2002).

Richard J. Norby, an environmental scientist at the Oak Ridge National Laboratory, said the Stanford study "is a surprise." "We don't really understand the responses of the plants [in the study]," said Norby, who is doing similar research in his lab. "I think this challenges some of our assumptions about global climate change." Norby said some conclusions have been based on studies that tested one or two elements of global change in laboratories. "What happens when all these conditions change at the same time is much more difficult to interpret," he stated (Recar 2002).

According to another study, "Well-watered and fertilized citrus trees clearly grow better when exposed to high levels of carbon dioxide, but the effect on forest trees is uncertain because most forests have limited supplies of other resources needed for growth" (Davidson and Hirsch 2001, 432). Trees fed additional nitrogen after natural supplies ran low continued to grow more quickly in a carbon dioxide—enriched atmosphere. "This gain was even larger at the poor site . . . than at the moderate site," another study suggested. "Here," conclude R. Oren and colleagues, "we present evidence that estimates of increases in carbon sequestration of forests, which is expected to partially compensate for increasing CO_2 in the atmosphere, are unduly optimistic" (Oren et al. 2001, 469). After an initial spurt of growth, trees exposed to elevated levels of carbon dioxide develop more slowly and do not absorb as much carbon dioxide from the atmosphere, according to Oren's work. This study suggests that reliance on forests to remove carbon dioxide from the atmosphere, a major part of the Kyoto Protocol, may not work. Planting trees alone will not substitute for reducing carbon dioxide emissions. As plants grow, they use nutrients in surrounding soils. As soils are depleted (at rates that increase with enhanced carbon dioxide), growth and ability to sequester carbon dioxide slow dramatically.

Bruce A. Hungate led a study of carbon dioxide "enrichment" in an oak woodland that indicated increasing nitrogen fixation during the first year; however, the effect declined in the second year and disappeared by the end of the third year. From the fourth through the seventh year of treatment, carbon dioxide enrichment consistently depressed nitrogen fixation. Clearly, the relationship between carbon dioxide level and nitrogen fixation is not simple. Hungate and colleagues found that "reduced availability of the micronutrient molybdenum, a key constituent of nitrogenase, best explains this reduction in N[itrogen] fixation. Our results demonstrate how multiple element interactions can influence ecosystem responses to atmospheric change" (Hungate et al. 2004, 1291).

In another study of the effects of elevated carbon dioxide levels on plant growth, Deborah Clark and colleagues commented that

> trees' annual diameter increments in this 16-year period (1984– 2000) were negatively correlated with annual means of daily minimum temperatures. . . . Strong reductions in tree growth and large inferred releases of CO_2 to the atmosphere occurred during the record-hot 1997–1998 El Niño. These and other recent

findings are consistent with decreased net primary production in tropical forests in the warmer years of the last two decades. . . . Such a sensitivity of tropical forest productivity to on-going climate change would accelerate the rate of atmospheric CO_2 accumulation. (Clark et al. 2003, 5852)

Clark, in a news report, mentioned, "We became interested in this when our long-term measurements of tree growth had come to cover several years, and we could see that tree growth varied greatly from year to year. We realized that something about yearly differences in weather patterns must be doing this, and we started to focus on what that might be" (Choi 2003). The scientists examined the annual growth of six tree species in an old-growth rainforest at the La Selva Biological Station in Costa Rica. They measured the trunk diameter of 164 adult tree species between 1984 and 2000 in an area the size of about 700 football fields. "To measure some of these trees required carrying two to three ladders cross-country through the forest in order to be able to measure the tree trunk diameter, often 3 to 10 feet above the ground to get above the large buttresses that protrude from many tropical trees," Clark said (Choi 2003).

When Clark and her team matched tree growth with local temperature readings, they found that the trees often were stunted during the hottest years—most notably during the record-warm 1997–1998 El Niño. During warmer years, atmospheric gas samples revealed that tropical regions as a whole released more carbon dioxide than they absorbed. In other words, tropical forests, sometimes called the "lungs of the world" for their capacity to produce oxygen, under some conditions actually became net emitters of carbon dioxide. "There is now an urgent need for studies to see if what we found at La Selva is occurring generally across tropical forests," Clark said. "If the patterns we have found prove to be general, it would mean that the rate of global warming will be much greater than what has been expected based on human fossil fuel use alone." She added, "No one knows what the optimum temperature is for photosynthesis of a tropical forest. This is clearly a burning question now" (Choi 2003).

"ENHANCED" CARBON DIOXIDE AND CROPS: MORE QUANTITY, LESS QUALITY

Increased levels of carbon dioxide in the atmosphere may accelerate growth of some crops but, at the same time, reduce their nutritional quality, according to a study published in *The New Phytologist* (Jablonski, Wang, and Curtis 2002, 9–26). Peter Curtis, a professor of evolution, ecology, and biology at Ohio State University and a coauthor of the study said, "If you're looking for a positive spin on rising CO_2 levels, it's that agricultural production in some areas is bound to increase. Crops have higher yields when more CO_2 is available, even if growing conditions aren't perfect. But there's a tradeoff between quantity and quality. While crops may be more productive, the resulting produce will be of lower nutritional quality" ("More Carbon" 2002).

Plants produce more seeds at higher carbon dioxide levels, but the seeds often contain less nitrogen. "The quality of the food produced by the plant decreases, so you've got to eat more of it to get the same benefits," Curtis said. "Nitrogen is a critical component for building protein in animals, and much of the grain grown in the United States is fed to livestock. Under the rising CO_2 scenario, livestock and humans would have to increase their intake of plants to compensate for the loss" ("More Carbon" 2002).

Curtis and colleagues conducted a meta-analysis in which they combined data from a large number of studies and then summarized for common trends. The studies the researchers reviewed, published between 1983 and 2000, included information on crop and wild plant species' reproductive responses to the doubled atmospheric carbon dioxide levels that are predicted to occur by the end of the twenty-first century. The researchers' analysis documented eight ways that plants respond to higher carbon dioxide levels: number of flowers, number of fruits, fruit weight, number of seeds, total seed weight, individual seed weight, the amount of nitrogen contained in seeds, and a plant's "reproductive allocation"—a measurement of a plant's capacity to reproduce ("More Carbon" 2002).

Food crops were influenced by varying levels of carbon dioxide to a much greater degree than were wild plants, according to the study. "Wild plants are constrained by what they can do with increased CO_2," Curtis explained. "They may use it for survival and defense rather than to boost reproduction. Agricultural crops, on the other hand, are protected from pests and diseases, so they have the luxury of using extra CO_2 to enhance reproduction" ("More Carbon" 2002).

Crops responded differently to increased carbon dioxide levels. Rice seemed to be the most responsive, for its seed production increased an average of 42 percent as carbon dioxide levels doubled. Soybeans followed with a 20 percent increase in seed, whereas wheat increased 15 percent and corn, 5 percent. Nitrogen levels decreased by an average of 14 percent across all plants except for cultivated legumes such as peas and soybeans ("More Carbon" 2002).

CLIMATIC WARMING MAY REDUCE RICE YIELDS

Even the modest temperature increases anticipated by some climate models could be enough to reduce rice yields significantly over the next century. Researchers at the University of Florida tested several varieties of rice, growing them in chambers that simulated various temperature-change situations. They found that, although the rice plants flourished no matter what the temperature, yields of grains declined precipitously as temperatures increased. Modest temperature increases, the researchers say, could reduce rice yields by 20 to 40 percent by 2100, whereas the larger increases predicted by more dire forecasts could cut rice production essentially to zero (Fountain 2000, F-5).

In another study, researchers from China, the United States, and the Philippines analyzed weather data from the International Rice Research Institute Farm in Los Banos, Philippines (near Manila), from 1979–2003 and compared these records with rice yields at the same location from 1992–2003. They found that annual mean maximum and minimum temperatures at the location increased by 0.35 degrees C and 1.13 degrees C, respectively, between 1979 and 2003; grain yield was found

Planting rice seedlings. Courtesy of Corbis.

to decline by 10 percent for each one degree rise in growing-season minimum temperatures during the dry cropping season (January through April). "This report," they wrote in the *Proceedings of the National Academy of Sciences,* "provides a direct evidence of decreased rice yields from increased nighttime temperature associated with global warming" (Peng et al. 2004, 9971). Globally, nighttime temperatures have increased faster than daytime readings because greenhouse gases tend to trap more heat radiating from the ground, reducing cooling (Fountain 2004, F-1).

Kenneth G. Cassman of the University of Nebraska, who participated in this study, said that researchers are working to determine the cause of the yield reduction, but they speculate that it is because the hotter nights make the plants work harder just to maintain themselves, diverting energy from growth. "If you think about it, world records for the marathon occur at cooler temperatures because it takes much more energy to maintain yourself when running at high temperatures. A similar phenomenon occurs with plants," he stated (Schmid 2004). Tim L. Setter, a professor of soil, crop, and atmospheric science at Cornell University (who did not take part in this study), commented that higher nighttime temperatures "could consume carbohydrates in a nonproductive way, and by reducing the reserves of carbohydrates, particularly at time of flowering and early grain filling, would decrease the number of kernels that would be set" (Schmid 2004).

CARBON DIOXIDE "ENRICHMENT" MAY FAVOR FASTER-GROWING PLANTS

Rising levels of carbon dioxide may favor faster-growing plants over those with slower growth rates. Referring to a set of experiments involving loblolly pines, a team of scientists wrote in *Science*: "We determined the reproductive response of 19-year-old loblolly pine (*Pinus taeda*) to four years of carbon dioxide enrichment (ambient concentration plus 200 microliters per liter) in an intact forest. After three years of CO_2 fumigation, trees were twice as likely to be reproductively mature and produced three times as many cones and seeds as trees at ambient CO_2 concentration" (LaDeau and Clark 2001, 95). Christian Korner, also writing in *Science*, asserted that such "plot studies" may not fully reflect carbon fluxes in real forests because they cannot take into account all the events in the life of a forest. For example, "In a fire, carbon fixed over a period of 50 to 300 years [the lifetimes of most trees], may be emitted within a few hours" (Korner 2003, 1242).

The carbon dioxide level LaDeau and Clark used to fumigate their pines will be typical of the ambient atmosphere by the year 2050, if present increases in greenhouse gases continue. The research suggests that faster-growing plants, such as pines, may have an advantage over slower-growing broadleaf trees as carbon dioxide levels rise. Another report on the same subject, also in *Science*, speculated that, "because CO_2 is a plant nutrient as well as a greenhouse gas, some researchers argue that faster-growing trees will absorb and sequester increasing amounts of CO_2, making it unnecessary to impose new controls on the gas" (Tangley 2001, 36). This view has been adopted with enthusiasm by climate "skeptics" who believe that nature will correct any greenhouse gas imbalance created by humans' combustion of fossil fuels at increasing levels.

17 FLORA AND FAUNA ON THE BRITISH ISLES

WARMING LENGTHENS GROWING SEASONS IN EUROPE

A network of phenological observers monitors gardens stretching from the north of Norway southeastward to Macedonia, as well as from Valentia Observatory's phenological garden in County Kerry, Ireland, eastward to Finland, using clones of a single sample of each species bred in Germany when the first gardens were established. Phenology, "the timing of seasonal activities of animals and plants" (Walther et al. 2002, 389), has been used to assemble a forty-year record that was analyzed by two German scientists, Annette Menzel and Peter Fabian, from the University of Munich. Observations gathered with this system indicate that spring events, including leaf unfolding, have advanced by about six days during this forty-year period, whereas autumn events such as leaf coloring now occur, on average, 4.8 days later. Therefore, the average growing season in Europe has lengthened by 10.8 days since the early 1960s (McWilliams 2001, 26).

These changes can be attributed almost entirely to changes in average temperatures, especially in winter. Other variables, including soil composition, water supply, biological factors, and the general surroundings, have as far as possible been kept identical. The results agree with a similar study carried out in the United States. That study concluded that during three decades the average first leaf date had moved progressively earlier each year, advancing from around May 4 in 1965 to April 23 by the early 1990s. According to one observer, "Both studies confirm what we already know—that average winter temperatures in both regions have been rising slightly, but significantly, throughout the period concerned. But more importantly, the results establish phenology as an important tool for the monitoring of global warming" (McWilliams 2001, 26).

During 2001 and 2002, British phenologists gathered information at various locations in the British Isles, "showing that the arrival of spring is no longer constrained to its traditional March slot and the end of autumn now continues well beyond late October." The phenologists' studies, taken collectively, indicated that "higher-than-average temperatures from January to April [of 2002] led to almost every characteristic of this year's spring occurring up to three weeks earlier than in 2001. On average, insects such as bees and butterflies, were three weeks early, while plants flowered two weeks ahead and many birds, including the turtle dove, arrived a week earlier than usual" (Hann 2002).

According to these studies, other notable early spring signs included:

- the call of the cuckoo—five days earlier;
- hazel flowering—twenty-three days earlier;
- emergence of snowdrops—seven days earlier;
- the first shimmering bluebell carpets—sixteen days earlier;
- alder leafing—thirteen days earlier;
- hawthorn leafing—seventeen days earlier.

Changes in the autumn included:

- beech leaf coloring—twelve days later;
- oak leaf coloring—nine days later;
- beech leaf fall—twelve days later;
- oak leaf fall—five days later; and, amazingly,
- "some people in milder parts of the United Kingdom were condemned to cutting their lawns all year-round" (Hann 2002).

Great Britain's Woodland Trust recruited 13,500 gardeners and wildlife watchers throughout the United Kingdom to monitor the first signs of spring early in 2004. A volunteer in Lochgilphead, Scotland, reported the first frogspawn on February 1, six weeks before its expected occurrence. The report also confirmed the earliest recorded sighting of a bumblebee in Scotland, on February 15. Ten more sightings were recorded during the same month. Blackbirds were observed building nests on January 8 near Kircudbright, two or three months ahead of usual. In London, during the 1920s and 1930s, bumblebees were never sighted before the first week of February. In 2003, one was sighted in Isleworth, West London, on December 23. Another was noted on Christmas Eve in Devon, lured out by mild temperatures. During the second week of February 2004, however, many of these "early wakers" were killed by a hard freeze, snow, and ice (Derbyshire 2004, 15).

The phenologists concluded that "interconnected and often complex relationships between trees, insects and animals in woods are also affected by early spring arrivals. For example, results from this spring [2002] show that some synchronous relationships between birds, insects and plants could become disturbed. Crucially, for many species, this could have serious implications for their future survival. As ancient woods become even more fragmented and isolated some characteristic plants and animals may face further threats as the climate changes" (Hann 2002).

EARLIER FIRST FLOWERING DATES IN ENGLAND

Meanwhile, a study of the first flowering date of 385 British plants found that about 200 species flowered an average of fifteen days earlier in the 1990s than in previous decades. The study, by Alastair Fitter of York University and his father, Richard, both distinguished naturalists, claimed to have found "the strongest biological signal yet of climate change" ("Warming Climate Pushes" 2002). The Fitters have been keeping track of blooming dates for half a century, long before global warming was a widespread concern. The senior Fitter was eighty-nine years of age in 2002. Fitter senior, who lives in Cambridgeshire, is one of Britain's best-known naturalists, the author of dozens of books on flowers and birds.

He began his recordkeeping after moving to Chinnor in Oxfordshire in 1953, when his son was five years of age.

A. H. Fitter and R.S.R. Fitter reported in *Science*: "The average first-flowering date of 385 British plant species has advanced by 4.5 days during the past decade compared with the previous four decades; 16 per cent of species flowered significantly earlier in the 1990s than previously, with an average advancement of 15 days in a decade. Ten species flowered significantly later in the 1990s than previously. These data reveal the strongest biological signal yet of climatic change" (Fitter and Fitter 2002, 1689). The Fitters also raised the possibility that climate change could alter evolution as different plants move into synchronous flowering patterns, raising the chances that they will breed natural hybrids (p. 1691).

Spring flowering species were the most responsive to the change, with white dead nettle flowering fifty-five days earlier during the 1990s than during four previous decades. From 1954–1990, its average first flowering date was March 18, but from 1991–2000, the plant bloomed on or about January 23. Ivy-leafed toadflax flowered five weeks earlier, and hornbeam, four weeks earlier, as did hairy bittercress. Lesser periwinkle flowered twenty-five days earlier, and lesser celandine, three weeks earlier. The opium poppy flowered twenty days earlier during the 1990s than its 1954–1990 average, according to the Fitters (Meek [May 31] 2002, 6). "Not everything is flowering earlier—otherwise you would just say that the beginning of spring has moved forward eight weeks," Fitter said ("Warming Climate Pushes" 2002). "I did an analysis of his data eight or nine years ago, up to 1990, thinking there might be some sign of climate warming, but there wasn't," the younger Fitter mentioned. "When we came to look at the last 10 years, out of the data jumped this incredible signal that it's really happening fast, and it's no coincidence that the 1990s were the warmest decade on record" (Meek [May 31] 2002, 6).

PROSPECTIVE PLANT EXTINCTIONS IN SCOTLAND

A computer modeling program called MONARCH (Modeling Natural Resource Responses to Climate Change), published in late

2001, asserted that climate change during the next fifty years could drive dozens of plant, bird, and animal species to extinction in Scotland. At highest risk from rising temperatures caused by global warming are two birds, the capercaillie and red-throated diver, and the mountain ringlet butterfly. The capercaillie is likely to lose 99 percent of suitable habitat, whereas the snow bunting will lose about a third of its climate area by 2020 and probably more than half by 2050.

Under similar conditions, the turtle dove, yellow wagtail, and reed warbler could expand their potential ranges, according to MON-ARCH, which forecasts that a number of other species could move into Scotland as the average annual temperature rises an estimated 13 degrees C during the years 2000–2050. "The predicted effects of rising temperatures can't be a good-news story," said Noranne Ellis, a climate change impact adviser with Scottish Natural Heritage. "But it's not all bad. We can also do something about the climate change which is taking place—everyone can play a part in reducing greenhouse gas emissions" (Maxwell 2001, 7). The MONARCH forecast assumed that the average temperature in Britain will rise by 3 degrees C in fifty years. "In the past decade the temperature and rainfall patterns have changed and we have seen, for example, the nuthatch move north for the first time to appear in Dumfries and Galloway. The holly blue butterfly has appeared and started breeding in Scotland," said Ellis (p. 7).

DECLINE OF THE CUCKOO IN ENGLAND

The cuckoo, England's herald of spring, a bird well known for the male's distinctive call (as well as for laying its eggs in nests assembled by other species), is disappearing from the British countryside. Cuckoo numbers have declined by 30 percent during the past thirty years in urban areas, while in woodland areas cuckoos have declined by as much as 60 percent. The cuckoo migrates to all parts of the country from winter feeding grounds in sub-Saharan Africa. David Marley of the Woodland Trust said that the cuckoo's preferred woodland habitat was particularly vulnerable to global warming, which could be affecting its breeding season and food supplies. "The cuckoo is one of the most amazing birds you can come across in Britain, but it is declining by a staggering amount and we are getting reports from across the country that people just aren't hearing it any more," Marley said. Other theories for the thirty-year slump in cuckoo populations include habitat loss and the spread of intensive farming practices. At its wintering places in Africa, cuckoos also may be suffering from indiscriminate use of agricultural chemicals as well as widespread drought (L. Smith 2002).

The British Trust for Ornithology (BTO) believes that cuckoos in Britain number between 12,000 and 24,000 pairs, down from 17,500–35,000 in 1970. The Woodland Trust has launched a study that will record cuckoo sightings and build data that will document the changing ecology, which may help explain why cuckoos are in sharp decline. New data will be added to that collected since 1726 to help monitor the impact of climate change. Volunteers will watch for cuckoos, as well as for other wildlife, including ladybirds, bumblebees, and swallows (L. Smith 2002).

Although some birds were declining in England as weather warmed, others were flourishing. For example, the number of wild parrots was rising rapidly by 2004, with 100,000 expected by the end of the decade. The parrots, which are large and aggressive birds, were competing with domestic species, such as starlings, jackdaws, and small owls, for food and territory (Prigg 2004, A-9).

WARMING, HABITAT DESTRUCTION, AND BUTTERFLIES IN BRITAIN

Some species of British butterflies that were expected to flourish in warmer temperatures have instead declined because of severe habitat destruction. Warm summers and mild winters since the 1970s should have attracted populations of butterflies that usually steer clear of the British Isles. According to a paper by M. S. Warren and colleagues in *Nature*, however, three-quarters of the butterfly species that might have expanded northward with warmer European weather have declined in numbers. The findings come from an analysis of 1.6 million butterfly sightings by 10,000 amateur naturalists between 1995 and 1999.

Chris Thomas of York University, who coordinated the study, said:

Most species of butterflies that reach the northern edge of their geographic ranges in Britain have declined over the past 30 years, even though the climate has warmed. This is surprising because climate warming is expected to increase the range of habitats these species can inhabit. . . . Our computer models show that climatically suitable areas are available for colonization, but most species have failed to exploit them either because they no longer contain suitable breeding sites or because breeding habitats are out of reach. (Derbyshire 2001, 13; Warren et al. 2001, 65)

Threats to butterflies of a warming climate in Britain also were described in a study compiled by British biologists and ecologists that was published by the Royal Society. According to this study, at least thirty of Britain's butterfly species face extinction or an alarming drop in numbers because they are failing to cope with the effects of global warming. These include some rare species, such as the large heath and purple empero, the numbers of which are expected to decline by as much as three-quarters. Many face eventual extinction as populations fall below replacement levels (Mason, Bailey, and London 2002, 12).

"There's no silver lining in this data," said Richard Fox, a coauthor of the study and spokesman for the Butterfly Conservation Society. "What we will see here is a retreat and, potentially, a mass extinction in the slightly longer term, of many of our familiar species." According to the report describing this work in the *London Independent*, "Over the past few years, Red Admirals, Orange-tips and Small Tortoiseshells have been seen earlier in the spring and surviving for several weeks longer each autumn, suggesting that butterflies would generally benefit from climate change." However, the researchers, led by Jane Hill, a biologist at York University, noticed that many more butterflies had failed to spread during the 1990s, even though average temperatures were beginning to rise (Mason, Bailey, and London 2002, 12). A few species, such as the ringlet and the marbled white have been prospering, moving northwards and further uphill as the summers became warmer during the 1990s, contrary to declining populations for many other species.

The butterfly study projects that, as warming accelerates in Britain, species living in northern England and Scotland, including the Western Isles, will lose two-thirds of their habitats. In southern England, butterfly habitat will decline by about a quarter. For some species, the future is particularly bleak. For example, the black hairstreak, a very rare species that has been confined mainly to the East Midlands, will lose at least half of its usual habitat. "We may get some butterflies from the south colonizing us, but what this paper shows is that our butterflies are going to become much, much rarer," Fox said (Mason, Bailey, and London 2002, 12).

INSECTS AND ARACHNIDS SPREAD NORTHWARD IN BRITAIN

Spiders, wasps, and similar species have been moving northward through Britain at unprecedented rates as temperatures rise. Insects with historic ranges in the northern reaches of England are moving into Scotland. Those previously resident in southern England have arrived in the north, as European spiders are increasingly invading Britain from the south and west. Global warming is being held responsible (Browne and Simons 2002, 8).

With no natural predators, scorpions have been spreading in England as temperatures rise. Two-inch-long, European yellow-tailed scorpions probably were brought to England from Italy or southern France during 1860 on ships docking at Sheerness in Kent. For many decades, a colony of about 1,000 hid in the brickwork of the dockyard. As temperatures warmed after the late 1980s, the scorpion colony grew to about 3,000 and spread into neighboring residential areas, where people have been advised to follow practices used in tropical areas: turn shoes upside down and shake them before wearing. Paul Hillyard, curator of arachnids at the Natural History Museum in London, said: "In the warmer temperatures these scorpions can flourish and people could start transporting them throughout Britain" (Ingham 2004, 40). European

A scorpion in an English brick wall. Photo by Ray Gabriel.

yellow-tailed scorpions also have been sighted in Harwich and Pinner and at the Ongar underground station in Essex.

According to one English observer, "The magnificent yellow and black wasp spider, *Argiope bruennichi*, which grows up to 50 millimeters (two inches) long, arrived in Hampshire from the Continent and now occurs as far north as Derbyshire" (Browne and Simons 2002, 8). Peter Harvey, head of the Spider Recording Scheme for the British Arachnological Society, reported that the wasp spider had spread rapidly during the five years ending in 2002. "It is almost certainly because of warmer and longer summers and autumns. It lays its eggs in autumn, so cold autumns are very bad for it" (p. 8).

Examples of insect migrations are plentiful in the British Isles:

The bee wolf, a wasp that feeds honeybees to its offspring, had until recently been confined to the Isle of Wight and East Anglia. However, it has spread rapidly in recent years and is now found as

The bee wolf wasp. Photo by Philip Penketh.

far north as Yorkshire. Arachnophobes in Scotland will be anxious to learn that the giant house spider, *Tegenaria gigantea*, which was once confined to England, has now reached Ullapool in the northwest Highlands. A species of jumping spider entirely new to Britain was recently discovered in London, having arrived from the Continent. A survey of Mile End Park in the East End of London has revealed a thriving colony of *Macaroeris nidicolens*, which is usually found in continental Europe, where it lives on small trees and bushes. Exactly how the spider arrived in Britain is unclear, although the young spiderlings could have crossed the

Channel using thread of gossamer blown on the wind. (Browne and Simons 2002, 8)

IRELAND'S CHANGING FLORA AND FAUNA

In Ireland, by 2001 swallows were arriving earlier, and the growing season for trees was steadily increasing, according to scientists at Trinity College of Dublin and National University of Ireland at Maynooth, who have been examining natural responses to temperature changes. Alison Donnelly of the Climate Change Research Centre, School of Botany, Trinity College, found that the beginning of the growing season, defined by the unfolding of leaves, has been occurring earlier in the spring. They also have noted that the end of the growing season, the time when the leaves fall, is occurring later in autumn, both of which point to the influence of rising global temperatures. Donnelly stated that, although records of the dates that swallows arrive on the east coast of Ireland only extend back to the 1980s, they nevertheless provide some evidence of the swallows' changing patterns of migration (Fahy 2001). Ireland's Environmental Protection Agency issued a report in late November 2002 titled "Climate Change Indicators for Ireland," which said that temperatures there have been rising 0.25 degrees C per decade, with increases accelerating during the 1990s. The number of frosty days had declined (Hogan 2002).

Hummingbird hawk moths, usually resident in southern Europe, have started appearing in Ireland, according to broadcaster and naturalist Eanna Ni Lamhna. Often mistaken for hummingbirds, the hawk moths hover as they swig nectar from flowers. Birdwatchers have also noticed an increase in the number of exotic birds. The little egret, a smaller version of the heron, has taken to nesting along the coast of Cork and Waterford. These birds, which are usually found in lands bordering the Mediterranean Sea, first started nesting in Ireland in 1997, and there are thought to be fifty pairs resident in this country. Other migrant species from southern climes that have recently appeared in Ireland include the pied flycatcher, the bearded tit, and the Mediterranean gull (Meagher 2004).

In the oceans surrounding Ireland, jellyfish have thrived in warmer sea temperatures, and in some enclosed bays, cockles expired from the warming waters. Karin Dubsky, the environmentalist who runs Coastwatch Ireland, stated that warming ocean temperatures may cause native oysters to be outbred by Pacific oysters being farmed along the coast. "We are worried that Pacific Oysters might thrive here as a result of the increase in sea temperatures. They could outbreed the Irish oyster in the same way that the grey squirrel knocked out the red squirrel" (Meagher 2004).

18 FLORA AND FAUNA: A WORLDWIDE SURVEY

Accelerated warming of the atmosphere is altering flora and fauna around the world. As the pace of warming increases, reports of its effects have become more numerous. Among the most notable effects have been increasing losses of pine forests in western North America to beetles and other pests. Entire regions have been killed from Utah to Alaska. Warmth accelerates the beetles' reproductive cycles. Other species have become more susceptible to infectious diseases. Warming has affected bird migrations; some have starved as migration patterns change but location of food sources does not. Some animals migrate upslope on warming mountains until they run out of elevation. The character of tropical forests has been changing as well, as certain species adapt to warming and others die.

RISING CARBON DIOXIDE LEVELS, LAND-USE CHANGES, AND TROPICAL FORESTS

University of Missouri scientist Deborah Clark, who works at the La Selva Biological Station in Costa Rica, reported at the annual meeting of the Association for Tropical Biology in Panama City, Panama, during the first week of August 2002 that tropical forests may soon become carbon dioxide generators rather than carbon sinks. She said that data from La Selva "show a strong negative correlation between tree growth and higher temperatures. Temperatures experienced by canopy leaves may be close to the point at which respiration exceeds photosynthesis so that net production of carbon dioxide results." As reported by the Environment News Service, she said that "positive feedback between higher temperatures and CO_2 production by tropical forests could be catastrophic by resulting in accelerated increase in global CO_2 levels" ("Global Warming Is Changing" 2002).

The devastation of tropical forests is not being caused by climate change alone. Deforestation also plays a major role. James (Bud) Alcock, professor of environmental sciences at Pennsylvania State University, has developed a mathematical model to study the effects of human-driven deforestation that forecasts the demise of the Amazon rainforest within forty to fifty years, given present-day deforestation rates of about 1 percent per year. Alcock stated that his model shows that "without immediate and aggressive action to change current agricultural, mining, and logging practices," the 2 million square mile Amazon rainforest could pass "the point of no return" in ten to fifteen years ("Deforestation Could Push" 2001). This model indicates that the rainforest could essentially disappear within 40–50 years, whereas other estimates give the area 75–100 years, given "business as usual."

"Because of the way tropical rainforests work, they are dependent on trees to return water to the air," said Alcock, noting that the size of the Amazon River Basin's forests has already been reduced by about 25 percent. "This interdependence of climate and forest means risks to the forests are much closer at hand than what we might expect." A healthy forest holds a large proportion of precipitation and returns it to the atmosphere so it can be recycled in a process called "evapotranspiration." Without a healthy base of vegetation, water runoff occurs at a higher rate and creates the potential for destruction of rainforest ecosystems. Less rain also could mean more frequent forest fires, further threatening the balance of the rainforest as well as the animal life it supports, Alcock said ("Deforestation Could Push" 2001).

By the early 1990s, Brazilian government figures (publication of which was delayed almost a decade because of political pressures) indicated that human activity was turning the Amazon valley from the "lungs of the world" into a net source of greenhouse gas emissions. A report by Claudio Dantas, "Brazil: A Major Air Polluter," published by the Brazilian newspaper *Correio Braziliense*, indicated that the main factor in this change was the burning initiated by settlers turning forests into farming areas. Largely because of deforestation, Brazil ranks among the world's top ten greenhouse gas sources. Deforestation is so pervasive in Brazil that it represents 70 percent of that nation's greenhouse gases, dwarfing the contributions of industry and transportation ("Amazon Deforestation" 2004).

Marcelo Furtado, campaign director at the environmentalist organization Greenpeace, says: "No government had the ability and political will to combat deforestation in Brazil. Considering that data from 1994 already showed a worrisome situation, it is obvious that things are much worse today." According to Furtado, deforestation reportedly increased 40 percent during the past decade. "We are reaching the point of 25,000 to 30,000 square kilometers of forest destruction. Urgent action must be taken," Furtado warned ("Amazon Deforestation" 2004). According to the Brazilian government, deforestation increased by 28 percent from 2002–2003 alone.

Mark Maslin and Stephen J. Burns have assembled a "moisture history" of the Amazon basin that indicates the area has been much drier in some past epochs than today: "[T]he Amazon River outflow history for the past 14,000 years . . . shows that the Amazon Basin was extremely dry during the Younger Dryas, with the discharge reduced by at least 40 per cent compared with that of today. After the Younger Dryas, a meltwater-driven discharge event was followed by a steady increase in the Amazon Basin effective moisture throughout the Holocene" (Maslin and Burns 2000, 2285).

During 2005 a severe drought spread through the Amazon Valley at the same time that new evidence was being assembled, indicating that damage from logging had been 60 to 123 percent more than previously reported. "We think this [additional logging] adds 25 percent more carbon dioxide to the atmosphere" from the Amazon than previously estimated, said Michael Keller, an ecologist with the U.S. Department of Agriculture's Forest Service, and coauthor of an Amazon logging inventory published in *Science* (Naik and Samor 2005, A-12; Asner et al. 2005, 480–481).

This new study differed from others that measured only the clear-cutting of large forest areas. The study by Asner and colleagues included these measures of deforestation and added trees cut selectively, while much of surrounding forest was left standing in five Brazilian states (Mato Grosso, Para, Rondonia, Roraima, and Acre), which account for more than 90 percent of deforestation in the Brazilian Amazon (Asner et al. 2005, 480). In addition, the Amazon Valley's worst drought in about forty years was causing several tributaries to evaporate, probably contributing even more carbon dioxide via wildfires.

The drought was, in some areas, the worst since record-keeping began a century ago. Some scientists asserted that the drought was most likely a result, at least partially, of a rise in water temperatures in the tropical Atlantic Ocean that also played a role in spawning Hurricane Katrina and other devastating storms during the 2004 and 2005 hurricane seasons. If global warming is involved, this drought may be only an early indication of a new weather regime in the Amazon Valley, which holds nearly a quarter of the world's fresh water.

"The Amazon is a kind of canary-in-a-coal-mine situation," said Daniel C. Nepstad, a senior scientist at the Woods Hole Research Center in Massachusetts and the Amazon Institute of Ecological Research in Belem (Rohter 2005). "A warmer Atlantic not only helps give more energy to hurricanes, it also aids in evaporating air," said Luiz Gylvan Meira, a climate specialist at the Institute for Advanced Studies at the University of São Paulo. "But when that air rises over the oceans in one region, it eventually has to come down somewhere else, thousands of miles away. In this case, it came down in the western Amazon, blocking the formation of clouds that would bring rain to the headwaters of the rivers that feed the Amazon" (Rohter 2005).

In Acre State in western Brazil, parched trees turned to tinder, and the number of forest fires recorded during 2005 tripled to nearly 1,500 by September, compared with a year earlier. The resulting smoke, which may itself have intensified the drought by impeding the formation of storm clouds, was so thick on some days that residents took to wearing masks when they went outdoors (Rohter 2005). "Because droughts remain registered in the soil for up to four years, the situation is still very critical and precarious, and will remain so," Nepstad said. Where there are "forests already teetering on the edge," he added, the prospect of "massive tree mortality and greater susceptibility to fire" must be considered (Rohter 2005).

F. Siegery and colleagues, writing in *Nature*, surveyed damage to tropical rainforests worldwide by fires during the 1997–1998 El Niño (ENSO) episode. They found that selective logging "may lead to an increased susceptibility of forests to fire" (Siegert et al. 2001, 437). This scientific team tested this hypothesis on East Kalimantan (Borneo), the part of Indonesia that suffered the worst fires, and found that "forest fires primarily affected recently logged forests; primary forests or those logged long ago were less affected" (p. 437). Recently logged forests contain an abundance of flammable dead wood, causing fires to spread more rapidly.

In addition, rising levels of carbon dioxide are causing a dramatic change in the composition of tree species in the Amazonian forest. A rising level of carbon dioxide is causing some tree species to grow faster and dominate forests. In turn, other species are being forced into decline. The rate at which trees absorb carbon dioxide and convert it into growth depends on the species. These factors help determine which species come to dominate in the forest and how useful the forest is as a "sink" that can soak up the pollution ("Carbon Pollution" 2004).

A U.S. and Brazilian team of biologists, led by William Laurance of the Smithsonian Tropical Research Institute in Panama and the National Institute for Amazonian Research in Manaus, Brazil, described the long-term effects of rising carbon dioxide levels on virgin Amazonian forests. They marked out eighteen one-hectare (2.5-acre) plots in central Amazonia, tagged nearly 13,700 trees with a trunk diameter of more than ten centimeters (4.5 inches), and monitored the growth and each species' population for twenty years. Of the 115 most abundant species, twenty-seven showed spectacular changes in population density and "basal area" (the amount of land occupied by the trunk of that species), which is a reliable indicator of biomass. Thirteen species gained in population density, and fourteen declined; fourteen species occupied a greater portion of the land, while thirteen species retreated ("Carbon Pollution" 2004).

The scientists found that "genera of faster-growing trees, including many canopy and emergent species, are increasing in dominance or density, whereas genera of slower-growing trees, including many sub-canopy species, are declining. Rising atmospheric CO2 concentrations may explain these changes" (Laurance et al. 2004, 171). "These changes could have both local and global consequences. Undisturbed Amazonian forests appear to be functioning as an important carbon sink, helping to slow global warming, but pervasive changes in tree communities could modify this effect" (p. 174).

The big winners as carbon dioxide levels rise have been spindly canopy-level trees and shrubs, such as the manbarklak, sclerobium, and parkia, which are fast-growing, with light-density wood. The losers have been slow-growing, dense tropical hardwoods, such as the croton and oenocarpus that live in the dark forest interior. These slow-growing trees are by far the biggest absorbers of carbon. They are the species that make the Amazon valley a vital carbon sink ("Carbon Pollution" 2004).

Human-initiated land-use changes have also been accelerating the Amazon's transition away from its role as a major carbon sink. Because jungle has been razed for cattle pasture, crops, logging, highways, and human settlements at an increasingly faster rate, what was once a large carbon sink may be turning into a net source of carbon. Scientists at the National Institute for Amazon Research in Manaus have estimated that carbon emissions in Brazil may have risen by as much as 50 percent since 1990. According to their calculations, land-use changes by 2002 produced annual emissions of 400 million tons of greenhouse gases, much more than the 90 million tons generated annually by use of fossil fuels in Brazil. "Right now, we cannot provide a definitive answer to the question of whether the Amazon is source or sink," Flavio Luizão, president of the International Scientific Committee of the LBA, said. "But in another three or four years, I think we will be able to reach a consensus" (Rohter 2003).

WARMING, DEFORESTATION, AND THE
DEVASTATION OF MOUNTAIN HABITATS

Tropical mountain forests depend on predictable, frequent, and prolonged immersion in moisture-bearing clouds. The felling of up-wind lowland forests alters surface energy budgets in ways that influence dry-season cloud fields (Lawton et al. 2001, 584). Cloud forests form where mountains force trade winds to rise above condensation level, the point of orographic cloud formation. "We all thought we were doing a great job of protecting mountain forests [in Costa Rica]," said Robert O. Lawton, a tropical forest ecologist at the University of Alabama in Huntsville, Alabama. "Now we're seeing that deforestation outside our mountain range, out of our control, can have a big impact" (Yoon 2001, F-5).

The clearing of forests, sometimes many miles from the mountains, alters this pattern, often raising the elevation at which clouds form. Lawton and colleagues used LANDSAT and Geostationary Operational satellite images to measure such changes in the Monteverde cloud forests of Costa Rica. Their "simulations suggest that conversion of forest to pasture has a significant impact on cloud formation" (Lawton et al. 2001, 586). Patterns found in Costa Rica resemble those in other tropical areas, including parts of the Amazon valley. "These results suggest that current trends in tropical land use will force cloud forests upward, and they will thus decrease in area and become increasingly fragmented—and in many low mountains may disappear altogether" (Lawton et al. 2001, 587).

Deforestation's effects probably extend further than most observers have heretofore believed. "Mountain forests . . . may be affected by what's happening some distance away," said Lawton. Each year about 81,000 square miles of tropical forests are cleared, Hartshorn said (Polakovic 2001, A-1). Thus, the weather in the lush cloud forests of Costa Rica is changing because of land-use changes, including deforestation, many miles away. As trees on Costa Rica's coastal plains are removed and replaced by farms, roads, and settlements, less moisture evaporates from soil and plants, in turn reducing clouds around forested peaks sixty-five miles away. At risk is the Monteverde Cloud Forest Reserve, an ecosystem atop a Central American mountain spine, "a realm of moss and mist, the woodland in the clouds—a type of rain forest—[that] is home to more than 800 species of orchids and birds, as well as jaguar, ocelot and the Resplendent Quetzal, a plumed bird sacred to the Mayans." The same area also is a watershed that supplies farms, towns, and hydropower plants in the lowlands (Polakovic 2001, A-1).

These findings are consistent with similar localized weather changes observed in deforested parts of the Amazon basin. Scientists say that cloud forests in Madagascar, the Andes, and New Guinea also are at risk. According to this study, "These results suggest that current trends in tropical land use will force cloud forests upward and they will thus decrease in area and become increasingly fragmented and in many low mountains may disappear altogether" (Polakovic 2001, A-1). "It's incredibly ominous that over such a distance deforestation can alter clouds in mountains. This is a very serious concern," said Gary S. Hartshorn, president of the Organization for Tropical Studies, a consortium of rainforest researchers at Duke University. "This is confirmation of what we have predicted for a long time," said Stanford University ecologist Gretchen Daily. "The implications are very serious for the tropics and other parts of the world" (p. A-1).

Using data collected from satellites and computer models, scientists examined how forest clearing along the Caribbean coastline, where more than 80 percent of lowland forests have been cleared for farms and towns, influences weather downwind in the Cordillera de Tilaran mountain range. Evaporation from lowland vegetation is a principal source of moisture for the 4,000- to 5,000-foot mountains during the dry season of January to mid-May.

As Gary Polakovic explained in the *Los Angeles Times*:

The researchers found that the moisture content of the clouds over the mountains has declined by about half since intensive land clearing began in the 1950s. Also, the cleared land is warmer, pushing the base of clouds nearly a quarter of a mile higher on some days, meaning they pass over the mountain range dropping little moisture. In contrast, clouds were more abundant over forested lowlands just across the border in Nicaragua, where forest still blankets much of the coastal plain. (2001, A-1)

Tropical rainforests are typically the last areas colonized by people, a final refuge of biodiversity in places where lowlands have been cleared and developed. "Many cloud forest organisms have literally nowhere to go," said Dr. Nalini M. Nadkarni, an ecologist at Evergreen State College in Olympia, Washington. "They're stuck on an island of cloud forest. If you remove the cloud, it's curtains for them" (Yoon 2001, F-5).

Many watersheds fed by wet highland forests also are threatened. "We always knew that if you have a town and a mountain behind it, you protect the mountain forests to protect the water," said James O. Juvik, a tropical ecologist at the University of Hawaii at Hilo. "This says, even if you leave the cloud forest intact like good conservationists, if you clear the lowland forests, you can diminish the cloud forest and affect your water flow" (Yoon 2001, F-5).

Begun as a colony of expatriate North American Quakers, Monteverde has been a major attraction for American biologists, many of whom have devoted their careers to the flora and fauna of the region. Carol Kaesuk Yoon described the area in the *New York Times:* "A quick look at the forest's inhabitants makes clear . . . the record-breaking diversity of orchids, to creatures like the resplendent quetzal, a glittering green, kite-tailed bird truly worthy of its name, and the so-called singing mice that whistle and chirp like birds" (2001, F-5). In addition to its value as a shelter for plants and animals, Monteverde also has become a center for eco-tourism. Nearly 50,000 tourists visit each year to walk its many trails, "stopping for a bite to eat or perhaps picking up a golden toad mug or Monteverde T-shirt along the way." Scientists say that if the cloud forest falters, much business will be lost along with its unique species.

Deforestation and a warming environment may help explain the disappearance of *Bufo periglenes*, the golden toad of the Monteverde tropical rainforest, which was found nowhere else on Earth. No one knows how long the golden toad had lived in the cloud forest that runs along the Pacific coast of Costa Rica. By 1987, however, the level of mountain clouds had risen, reducing the frequency of mists during the dry season and probably playing a role in a massive population crash that affected most of the fifty species of frogs and toads in the forest. No fewer than twenty species became locally extinct (Moss 2001, 18).

According to Yoon, "A flashy orange creature last seen in the late 1980s, the golden toad has become an international symbol of the world's disappearing amphibians. Seeing how even pristine and protected forests such as those at Monteverde can lose their crucial mists and clouds, researchers say it becomes less mysterious how a water-loving creature like the golden toad could vanish even from a forest where every tree still stands" (2001, F-5).

By 2006, the role played by warming habitats in triggering disease among amphibians was becoming clearer, following studies of the Monteverde harlequin frog (*Atelopus sp.*), which vanished along with the golden toad (*Bufo periglenes*) during the late 1980s in the mountains of Costa Rica. According to analysis by J. Alan Pounds (resident scientist at the Tropical Science Center's Monteverde Cloud Forest Preserve in Costa Rica) and colleagues in *Nature* (2006, 161–167), an estimated 67 percent of roughly 110 species of *Atelopus*, which are endemic to the American tropics are now extinct, having been adfflicted by a pathogenic chytrid fungus (*Batrachochytrium dendrobatidis*). Pounds and colleagues concluded with more than 99 percent confidence that large-scale warming is a key factor in the disappearances. They found that rising temperatures at many highland localities have encouraged growth of *Batrachochytrium*, encouraging outbreaks. "With climate change promoting infectious disease and eroding biodiversity, the urgency of reducing greenhouse-gas concentrations is now undeniable," they concluded (Pounds et al. 2006, 161).

"Disease is the bullet killing frogs, but climate change is pulling the trigger," Pounds said. "Global warming is wreaking havoc on amphibians and will cause staggering losses of biodiversity if we don't do something first" (Eilperin 2006, A-1). The chytrid fungus kills frogs by growing on their skin, then attacking their epidermis and teeth, as well as by releasing a toxin. Higher temperatures allow more water vapor into the air, which forms a cloud cover that leads to cooler days and warmer nights. These conditions favor the fungus, which grows and reproduces best at temperatures between 63 and 77 degrees F. "There's a coherent pattern of disappearances, all the way from Costa Rica to Peru," Pounds said. "Here's a case where we can show that global warming is affecting outbreaks of this disease" (Eilperin 2006, A-1).

AMPHIBIANS THREATENED WORLDWIDE

In 2004, the first worldwide survey of 5,743 amphibian species (frogs, toads, and salamanders) indicated that one in every three species was in danger of extinction, many of them likely victims of an infectious fungus possibly caused by drought or global warming (Stokstad 2004, 391). These findings represented the combined research of more than 500 scientists from more than sixty countries. "The fact that one-third of amphibians are in a precipitous decline tells us that we are rapidly moving toward a potentially epidemic number of extinctions," said Achim Steiner, director-general of the World Conservation Union based in Geneva (Seabrook 2004, 1-C). "Amphibians are one of na- ture's best indicators of the overall health of our environment," said Whitfield Gibbons, a herpetologist at the University of Georgia's Sa- vannah River Ecology Laboratory (Stokstad 2004, 391).

Amphibians are more threatened and are declining more rapidly than birds or mammals. "The lack of conservation remedies for these poorly understood declines means that hundreds of amphibian species now face extinction," wrote the scientists who conducted the worldwide survey (Stuart et al. 2004, 1783–1786).

In North and South America, the Caribbean, and Australia, a major culprit appears to be the highly infectious fungal disease chytridiomy- cosis. New research shows that prolonged drought may cause outbreaks of the disease in some regions, although some scientists attribute the disease spread to global warming. Other threats include loss of habitat, acid rain, pesticides and herbicides, fertilizers, consumer demand for frog legs, and a depletion of the ozone layer that leaves the frogs' skin exposed to radiation. Gibbons said that the loss of wetlands and other habitat to development, agriculture, and other reasons may be the leading cause of amphibian declines in Georgia. Pollution also may be playing a significant role, he said. "It's hard to find a pristine stream anymore," he commented (Seabrook 2004, 1-C).

Warming and the Decline of Oregon's Western Toad

Global warming could be playing a role in the decline of the western toad in Oregon, according to a report that was among the first to link climatic change with amphibian die-offs in North America. The wave of toad population declines may be a harbinger of ecological chain reactions that result from rapid global warming. J. Alan Pounds, who has researched the decline of the golden toad in Costa Rica from his post with the Monteverde Cloud Forest Preserve and Tropical Science Center in Costa Rica, described the situation in Oregon: "In the crystal-clear waters surrounded by snow-capped peaks in the Cascade range, the jet-black [western toad] embryos are suffering devastating mortality. They develop normally for a few days, then turn white and die by the hundreds of thousands" (2001, 639).

Research by Joseph M. Kiesecker of Pennsylvania State University and colleagues indicates that the toads' fatal infection results from a complex sequence of events provoked by warming temperatures. These researchers wrote that "elevated sea-surface in this region [the western United States] since the mid-1970s, which have affected the climate over much of the world, could be the precursor for pathogen-mediated amphibian declines in many regions" (Kiesecker, Blaustein, and Belden 2001, 681). "Reductions in water depth due to altered precipitation patterns expose the embryos to damaging ultraviolet b (UV-B) radiation, thereby opening the door to lethal infection by a fungus, *Saprolegnia ferax*" (Pounds 2001, 639).

Depth of water influences the amount of UV-B radiation that reaches the toads' eggs; less water depth is related to the El Niño/La Niña cycle, which, Kiesecker and colleagues theorize, is related to climate change. Similar patterns have been detected in cases of mass frog and toad mortality from other parts of the world. In some areas, fungus-borne diseases afflict lizards as well as frogs and toads (Pounds 2001, 639–640).

Given what Kiesecker and others have found, Pounds asserts that "there is clearly a need for a rapid transition to cleaner energy sources if we are to avoid staggering losses of biodiversity" (p. 640).

Kiesecker's results support those of Pounds, who in 1999 reported that warmer and drier periods in the cloud forest atop the continental divide at Monte Verde were tied to El Niño events and had caused massive population reductions in more than twenty frog species, including the disappearance of the golden toad (Souder 2001).

SEA BIRDS STARVE AS WATERS WARM

The survival of sea birds on New Zealand's sub-Antarctic islands is directly related to supplies of krill, which are declining as waters warm. Populations of rockhopper penguins, a species with brilliant yellow eyebrows and that nests on the rocks of windblown Campbell Island, have declined by more than 95 percent since the 1940s. A breed of albatross called grey-headed mollymawks has declined by 84 percent. Muttonbird numbers are down by a third, and elephant seals, by about half, according to a report in the *New Zealand Herald* (Collins 2002).

A study by New Zealand's National Institute of Water and Atmospheric Research (NIWAR) said that the collapse of the rockhopper penguin population is due to the birds' inability to find enough food in a less-productive marine ecosystem. "The cause is reduced productivity of the ocean where they are feeding," said NIWAR scientist Paul Sagar. "They eat mainly krill but they do also eat small fish and squid." This research measured the amount of food available to the penguins in the sea during the preceding 120 years by analyzing feathers from rockhoppers alive today compared with those of forty-five museum specimens from the Antipodes and Campbell Island. Results indicated declining numbers of the phytoplankton in the rockhoppers' diet (Collins 2002).

Sagar said that further research is required to discern why phytoplankton populations have declined; some scientists assert that the declining populations result from water that has become too warm for a species that is adapted to the sub-Antarctic. "It could be a long-run natural cycle, with the possibility that polar water coming up from the south may not be coming as far north as it used to," Sagar said. "It could be that the position of the currents has changed so the productive areas have moved further south or north, away from where the penguins are feeding. If it's happening over such a long time, I wouldn't put it down specifically to global warming" (Collins 2002). He said that the study used the rockhoppers as an "indicator species," because it was likely that the same changes in climate and phytoplankton populations also were causing decline in other bird populations.

Another study, at New Zealand's Otago University, indicated that muttonbird numbers had dropped by a third on one of their major breeding islands. Conservation Department scientist Peter Moore, who studied the rockhoppers on Campbell Island in 1996, said that the rockhopper numbers declined from 1.6 million breeding pairs in 1942 to 103,000 pairs in 1985 and had continued to fall at a similar rate since. Sagar said there was evidence that some of the adult birds were feeding better-quality food to their chicks, while their own diet declined, but Moore said the orphaned chicks often died as well. During 1990 many surviving chicks from a yellow-eyed penguin colony were fed and reared by humans after their parents died. But after the chicks were released, 99 percent of them disappeared (Collins 2002).

Sea bird population declines have spanned the world, for similar reasons. In Scotland, for example, guillemots, arctic terns, kittiwakes, and other sea bird colonies on the Shetland and Orkney islands in 2004 experienced one of their worst breeding seasons in memory. The Royal Society for the Protection of Birds (RSPB) observed very few chicks on the breeding cliff ledges. Sandeels, the small fish on which the birds feed, have migrated northward, probably because of warming waters, placing the birds' traditional food supply largely out of reach, so they have had difficulty reproducing and feeding their young.

CLIMATE CHANGE AND BIRD MIGRATIONS: THE PIED FLYCATCHER

Dutch researchers have reported how one long-distance migratory bird responds to climate change. The researchers studied pied flycatchers, which spend winters in West Africa and return to woodlands in Holland each spring. Average temperatures in the spring territory have increased several degrees during the last twenty years. According to the researchers, higher spring temperatures mean that insects, the birds' food source, reach peak abundance earlier, providing birds that hatch at the same time with more food.

The researchers found that over the two decades, the pied fly-catcher's mean laying date has advanced by a little more than a week, as selection has favored earlier laying. The dates that the birds migrate have not changed, the researchers believe, because the birds' decision to leave their winter habitat is related to the amount of daylight, which is not affected by warming. The birds are arriving at the same time, but laying eggs earlier. The researchers suggest that this phenomenon may be contributing to declines of other long-distance migrants (Both and Visser 2001, 296; Fountain 2001, F-4).

BALTIMORE WITHOUT ORIOLES: ANTICIPATED BIRD EXTINCTIONS

Maryland's Baltimore orioles, which have been declining due to habitat loss for many years, could vanish altogether late in the twenty-first century due to changes in migration patterns strongly influenced by a warming climate. A study by the National Wildlife Federation and the American Bird Conservancy suggests "that the effects of global warming may be robbing Maryland and a half-dozen other states of an important piece of their heritage by hastening the departure of their state birds." The report stated that Earth's rising temperature "is already shifting songbird ranges, altering migration behavior and perhaps diminishing some species' ability to survive" (Pianin 2002, A-3).

Iowa and Washington State may lose the American goldfinch, and New Hampshire's purple finch could become an historical relic. California could lose the California quails, Massachusetts' black-capped chickadee may vanish, and Georgia could lose its brown thrasher (Pianin 2002, A-3).

The life cycle of the oriole and other birds is tied closely to weather patterns that are changing with general warming. Seasonal changes in weather patterns tell the birds when they should begin their long flights southward in the fall and back again in the spring. Temperature and precipitation also influence the timing and availability of flowers, seeds, and other food sources for the birds when they reach their destinations (Pianin 2002, A-3).

Peter Schultz, a global warming expert with the nonprofit National Research Council, cautioned that long-term forecasts of disruptions in bird migration patterns are difficult. "I would be surprised if the distribution of state birds is not changed down the road," he said. "But predicting precisely where they'll be 50 years from now is very difficult, if not impossible, with the current state of knowledge" (Pianin 2002, A-3).

Baltimore orioles (*Icterus galbula*) once were so numerous that the naturalist-painter John J. Audubon wrote about the delight of hearing "the melody resulting from thousands of musical voices that come from some neighboring tree" (Pianin 2002, A-3). The bird, a Maryland icon whose namesake was adopted by Baltimore's major league baseball team, was officially designated the state bird in 1947. Local legend maintains that George Calvert, the first baron of Baltimore, liked the oriole's bright orange plumage so much that he adopted its colors for his coat of arms.

Global warming is not the only danger to the oriole and other well-known birds. Their decline results also from diminishing breeding habitat and forests in North America (where orioles spend summers) and in Central and South America, where they fly for the winter. "Climate change on top of fragmented habitat is the straw that breaks the camel's back," said Patricia Glick, an expert on climate change with the National Wildlife Federation (Pianin 2002, A-3).

BRITISH COLUMBIA TUFTED PUFFINS AND CLIMATE CHANGE

Roughly 50,000 tufted puffins that spend summers on Triangle Island, thirty miles off the British Columbia coast, have been falling prey to periods of starvation because small changes in ocean temperature have driven away their food supply. Researchers said that the adult birds bring back far fewer sand-lance fish to their young in warm years. The sand-lance fish, the puffin's favored food, become less abundant in the waters around Triangle Island as water temperatures rise. Scientists believe the lack of food, coupled with self-preservation instincts in the adults, leads to abandonment of chicks.

"The difference between 1998 and 1999 was one of the most striking things I have ever seen in my career in ecology," said Doug Bertram, a marine bird specialist at the Canadian Wildlife Service, as he recalled how dead chicks littered the colony in 1998 (Munro 2003, A-12). "We show that the extreme variation in reproductive performance exhibited by tufted puffins (*fratercula cirrhata*) was related to changes in sea-surface temperatures both within and among seasons," the authors of a scientific study of the birds explained. Such changes in ocean temperatures "could precipitate changes in a variety of oceanic processes to affect marine species worldwide" (Gjerdrum et al. 2003, 9377).

Dismayed researchers watched helplessly as chicks dropped dead of starvation during several warm summers in the 1990s. In other years, such as 1999 when water was 1.5 degrees C cooler, the chicks thrived (Munro 2003, A-12). By 2003, puffin populations had returned to former levels as water temperatures cooled to near-average levels. The scientists worried, however, that the respite may be temporary if global warming causes ocean temperatures to rise again in coming years.

HIGH-ELEVATION PIKA POPULATIONS PLUNGE

Pikas—described by Usha Lee McFarling of the *Los Angeles Times* as "tennis ball-sized critters that whistle at passing hikers and scamper over loose, rocky slopes of the High Sierra and the Rocky Mountains, . . . a shy, flower-gathering mammal and longtime icon of the West's high peaks"—may fall victim to global warming. By 2003, they already had disappeared from nearly 30 percent of the areas where they were common in the early parts of the twentieth century. Pikas, which resemble hamsters, are biological relatives of rabbits. Over many

Pika. Courtesy of Getty Images/PhotoDisc.

millennia, they adapted to a degree of intense cold that is becoming rarer in the mountains. Living above 7,000 feet, the pikas cannot withstand heat. A comprehensive survey found that sites that have lost pikas were on average drier and warmer than sites where the animals remain, said Erik Beever, a U.S. Geological Survey biologist based in Corvallis, Oregon (McFarling 2003, A-17).

Beever indicated that such factors as cattle grazing and proximity to roads had some effect on the animals. Warmer and drier conditions in recent decades have been a major factor in their rapid disappearance, however. Earlier surveys in the United States had found pikas missing from much of their previous range. Work on pikas in the Yukon revealed that 80 percent of the animals died in some areas after unusually warm winters. Beever reported that the demise of the pika has been very rapid. He was most surprised to find groups of animals disappearing over decades, rather than in centuries or millennia as during climatic swings of the distant past (McFarling 2003, A-17).

THE PITCHER-PLANT MOSQUITO EVOLVES TO SUIT GLOBAL WARMING

Some creatures have shown an unusual ability to adapt quickly to warming habitats. University of Oregon researchers, for example, have been studying the pitcher-plant mosquito, a tiny, fragile species that seldom bothers people. This species has delayed the dates at which it breeds and develops, according to a report in the *Proceedings of the National Academy of Sciences*. The pitcher-plant mosquito bases its life cycle on the length of the day. When winter approaches, it is genetically programmed to hibernate. A subgroup within the population has slightly different genes that cause it to reproduce later in the season, however.

Global warming, with its longer growing seasons, eventually could lead this group to dominate the pitcher-plant mosquito population. Mosquitoes collected from Florida to Canada between 1972 and 1996 were found to have shifted their breeding and development patterns as temperatures have risen, according to a team led by William E. Bradshaw of the University of Oregon. Because this behavior is genetically programmed, the researchers conclude that the mosquito "represents an example of actual genetic differentiation of a seasonality trait that is consistent with an adaptive evolutionary (genetic) response to global warming.... Our results suggest that other species may be in the process of analogous evolutionary response" (Stein and Vedantam 2001, A-9). "This shift is detectable over a time interval as short as 5 years," representing "genetic differentiation of a seasonality trait that is consistent with adaptive evolutionary response to recent global warming" (Bradshaw and Holzapfel 2001, 14509).

RED SQUIRRELS' BREEDING SEASON ADVANCES IN THE YUKON

Birth dates of red squirrels in the Yukon advanced almost three weeks within a decade because of a warming habitat, according to scientists who assert that the change is so profound that it seems to have affected the creature's genetic makeup (Munro 2003, A-2). "It's a phenomenal change in such a short period of time," said Stan Boutin, a biologist at the University of Alberta, who tracked hundreds of squirrels in the southwest corner of the Yukon. "We feel we've found some of the first evidence indicating climate change is leading to evolution in animal populations," remarked Boutin, who worked with Andrew McAdam, also from the University of Alberta, Denis Reale from McGill University, and Dominique Berteaux from the Universite du Quebec (Ogle 2003, A-12). Such changes simulate what occurs when animal breeders select for certain traits that are gradually bred into the animals over several generations, he said. In this case, however, climate change is driving the evolutionary change among the squirrels through natural selection.

"It's typical Darwinian natural selection happening," Boutin said. "As the conditions get warmer in spring, it favors those individuals that carry genes that allow them to breed earlier" (Munro 2003, A-2). Boutin studied four generations of squirrels during a fifteen-year period at Kluane National Park. Their study, published in the March 2003 edition of the *Proceedings of the Royal Society of London* (270:591–596), is the first to suggest that warming has triggered genetic change.

The squirrels' breeding season has been markedly affected by warmer spring temperatures and an increased food supply from spruce trees, which produce more cones as temperatures warm. Pups are now born as early as March 1. "It's good news in the sense that we are really impressed by the adaptability of these characters and their ability to keep up with the change," said Boutin. "The bad news is that these rates of change are definitely very rapid. Our human activities are affecting organisms in a variety of ways, even to the point where we are causing evolution in them" (Munro 2003, A-2).

As this study was released during 2003, observers in British Columbia reported that a mild winter there was prompting animals to procreate several weeks earlier than usual. The first house sparrow hatchlings appeared in the city several weeks earlier than usual, said Stanley Park Ecology Society spokesman Robert Boelens. Boelens expected that other urban critters may be mating earlier than usual—everything from ducks to squirrels to swans to raccoons. "In fact," according to a

Vancouver Sun report, "squirrel love is definitely under way. . . . Several people around the Lower Mainland [near Vancouver's urban area] have reported significant squirrel nest-building activity weeks earlier than usual. The same behaviors have been reported for raccoons, skunks, rabbits, mice, crows, and rats" (Reed 2003, B-1).

WARMTH ALTERS THE SEX RATIO OF LOGGERHEAD SEA TURTLES

As many as 85 percent of loggerhead sea turtle hatchlings living on beaches in the southern United States are now female, a sex ratio caused

Loggerhead turtle, Rocktail Bay, Maputaland, South Africa. Courtesy of Getty Images/Digital Vision.

by a warming habitat that threatens this endangered species. A lack of males may cause the species to become extinct. "These turtles have very small gonads at this age and are difficult to identify," said Jeanette Wyneken, an assistant professor of biological sciences at Florida Atlantic University, who is an expert on sea turtle anatomy and turtle conservation. According to a report by the Environment News Service, "The skewed sex ratios can arise because the temperature of the sand surrounding a turtle nest plays a strong role in determining the sex of turtles, with warmer temperatures favoring females" (Lazaroff 2002).

JAPANESE FRUITS SPREAD NORTHWARD

The goya, a bitter Japanese melon that looks like a knarly cucumber, until recently had been grown mainly on Okinawa. By the year 2000, however, with warming weather, farmers were cultivating the melon as far north as some parts of Honshu. The *unshu-mikan*, similar to a mandarin orange, has been threatened as warm weather has prompted trees to leak sap even in winter, an invitation to pests. The harvest of yellowtail, or *kan-buri*, a cold-water fish, has been declining rapidly in Japanese waters, while harvests of sardines, which have an affinity for warm water, have been increasing ("Bitter Pill" 2003).

WARMING, HUMAN ENCROACHMENT, AND REDUCTION OF THE GELADA BABOON

Human encroachment and warming temperatures in the highlands of Ethiopia are eliminating the habitat of the gelada baboon. Primate expert Chadden Hunter stated that "our research suggests that for each 2 degrees C. [3.6 F] degree increase in temperature, the gelada's lower limit for grazing will rise 500 meters [547 yards].... Global temperatures only need to rise a few degrees, and [their habitat] will be, in effect, lifted off the tops of the mountains" (Smucker 2001, 7). A Swiss study of the Ethiopian highlands has shown that the specific grasses on which the baboons feed have receded steadily upward during the past several hundred years. A separate study commissioned by the World Wide Fund for Nature last year cited the gelada baboon as one of several mammals most at risk from the effects of global warming.

American and Ethiopian primate experts concur that temperature shifts may soon destroy the high-altitude "islands of grass" on which the gelada survives. "Since the gelada survive at the physical limits of the landscape, the likelihood of future global warming raises serious questions over the species' survival," said Jacinta Beehner, an American baboon expert with long experience in Ethiopia (Smucker 2001, 7).

The gelada baboon. Courtesy of the Wildlife Conservation Society.

Theropithecus gelada is the last remaining species of its type. Hunter believes that the graminivorous (grass-eating) monkeys may have lost most of their close relatives to past episodes of climate change. "They are the last relic of a great dynasty, barely surviving in the Ethiopian highlands." The gelada, which scamper and leap through the highlands in "herds" of 600 or more, maintain a complex social structure that offers "irreplaceable insight into human social behavior," said Hunter (Smucker 2001, 7).

Although warming may provoke the baboons' demise within a century, "the most immediate threat to the gelada's survival comes from farming," commented Tesfaye Hundessa, manager of the Ethiopian Wildlife Conservation Organization. "They could well be entirely eliminated by humans before the global warming has time to take its toll over the coming decades" (Smucker 2001, 7).

WARMING AND THE SPREAD OF INFECTIOUS DISEASES AMONG WILDLIFE

Warming temperatures have been triggering epidemics of infectious diseases in wildlife around the world, ecologists and epidemiologists have warned. A scientific team's two-year study was the first comprehensively to analyze worldwide epidemics across entire plant and animal systems, on land and in the oceans. C. Drew Harvell of Cornell University's ecology and evolutionary biology department, lead author of the study, said: "What is most surprising is the fact that climate-sensitive outbreaks are happening with so many different types of pathogens—viruses, bacteria, fungi and parasites—as well as in such a wide range of hosts, including corals, oysters, terrestrial plants, birds, and humans" (Cookson 2002, 4). The spread of diseases, in most cases, is supported by increased vigor associated with warming and increased moisture that the team attributed to various "vectors," organisms that spread diseases such as mosquitoes, ticks, and rodents. "The accumulation of evidence has us extremely worried," said epidemiologist Andrew Dobson of Princeton University. "We share diseases with some of these species. The risk for humans is going up" (p. 4).

Harvell and colleagues found that "infectious diseases can cause rapid population declines or species extinctions. Many pathogens of terrestrial and marine taxa are sensitive to temperature, rainfall, and humidity, creating synergisms that could affect biodiversity." Climate warming can increase pathogen development and survival rates, disease transmission, and host susceptibility. Climate changes associated with El Niño events that may be related to global warming also have had a "detectable influence" on pathogens, including coral diseases, oyster pathogens, crop pathogens, Rift Valley fever, and human cholera (Harvell et al. 2002, 2158).

Harvell and colleagues reviewed potential consequences of temperature changes on infectious diseases and considered the hypothesis that climate warming will affect host-pathogen interactions by (1) increasing pathogen development rates, transmission, and number of generations per year; (2) relaxing overwintering restrictions on pathogen life cycles; and (3) modifying host susceptibility to infection (Harvell et al. 2002, 2158). The scientists anticipate that greater overwintering success of pathogens will likely increase the severity of many diseases. Because temperatures are expected to increase more in winter than in other seasons, this "population bottleneck" may be removed for many pathogens. "Several plant diseases are more severe after mild winters or during warmer temperatures, which suggests that directional climate warming will alter plant disease severity," they wrote (p. 2159). Harvell and colleagues cited an example of fungi infecting Mediterranean oak and the Dutch elm disease. "Warming," they wrote, "can decrease plant resistance to both fungi and viruses. Plant species that have faster growth rates in warmer climates also may experience increased disease severity, because higher host density increases the transmission of many pathogens" (p. 2159). Harvell and colleagues further stated that vector-borne human pathogens, such as malaria, Africa trypanosomiasis, Lyme disease, tick-borne encephalitis, yellow fever, plague, and dengue, have increased in incidence or geographic range in recent decades (p. 2160). Similar expansions have been noted in animal diseases.

Warmer summers and milder winters are encouraging diseases that threaten lions, cranes, vultures, and ferrets. "The global warming is also helping to spread tropical diseases to human habitations previously unaffected by such illnesses," according to Harvell and colleagues (Radford 2002, 7). "The number of similar increases in disease incidence is astonishing," said Richard Ostfeld, of the Institute for Ecosystem Studies in Millbrook, New York (p. 7). Warming is shrinking areas where pests are controlled by frosts and where cool periods can cut insect, parasite, and fungus pest populations by up to 99 percent. By one account, "In Hawaii, avian malaria has already wiped out native song birds living below an altitude of 1,400 meters [4,500 feet]" (p. 7).

The team reserved special attention for El Niño events. The scientists looked at the massive die-off of corals during the unusually warm El Niño year of 1998. Much of the coral had died from fungal and other diseases thriving in warmer seas. Damage was found to be worldwide. Oysters in Maine have been blighted by parasites normally restricted to more southerly waters. Lions in the Serengeti suffered canine distemper, and cranes, vultures, and even wild American ferrets had been hit by disease outbreaks. The monarch butterfly came under pressure from an exploding parasite population.

The scientists fear that the spread of tropical infections to temperate zones could bring calamity. Compared with other geographical regions, the tropics have a richer variety of species and fewer individuals in each species; diversity acts as a buffer against the spread of disease. The temperate zones have a smaller range of species with greater numbers in each, so pathogens moving into the temperate zones could affect a few common and abundant susceptible species (Radford 2002, 7). "Human destruction of biodiversity makes this a double whammy. It means we are exacerbating the problem," said Dobson. "We have to get serious about global change. It is not only going to be a warmer world, it is going to be a sicker world" (p. 7).

Almost everywhere the authors looked—on land and sea, in tropical climes and temperate ones—they found examples of plants and animals falling prey to disease as temperatures rose. "When you see the same pattern across so many organisms, you need to start taking the climate signal seriously," said Richard S. Ostfeld. "We don't want to be alarmist, but we are alarmed" (McFarling 2002, A-7). For example, Ostfeld cited malaria-carrying mosquitoes that "invade higher elevations, leading to more malaria deaths in mountain-dwelling Hawaiian songbirds. Ligurian Sea sponges that suffer heat stress in warmer waters are more vulnerable to fatal infection" (p. A-7).

The authors of this study also included some examples, paradoxically, in which warmer temperatures could lead to less disease. These include a number of diseases in fish, insects, and amphibians that spread only in cool conditions. Harvell said that her team could not yet make specific forecasts on how warming might alter ecosystems, but she said there was concern because "diseases are capable of reshaping whole communities very rapidly." She pointed to Caribbean corals, where many species that existed for 4,000 years have been wiped out by disease during the last fifteen years (McFarling 2002, A-7). "This is a truly disturbing panorama," Paul Epstein of Harvard Medical School's Center for Health and the Global Environment told Usha Lee McFarling of the *Los Angeles Times*. "I'm afraid we've underestimated the true costs of what climate change is doing to our environment and our society" (p. A-7).

BARK BEETLES SPREAD ACROSS THE U.S. WEST

Six years of intense drought and rising temperatures by 2004 were creating perfect conditions for bark beetle infestations across the U.S. West (see Volume 2, Part III: "Icemelt around the World," for information on pine beetle infestations in Alaska). By the fall of 2002, large swaths of evergreen forests in western Montana and the Idaho panhandle, as well as parts of California, Colorado, and Utah, had fallen victim to unusually large infestations of bark beetles, including the Douglas fir bark beetle, spruce beetle, and mountain pine beetle. The infestations were being encouraged by several factors: a warming trend, which allowed the beetles to multiply more quickly and reach higher altitudes; drought, which deprived trees of sap they would usually use to keep the beetles under control; and years of fire suppression, which increased the amount of elderly wood susceptible to attack. Beetles, attacking in "epic proportions," have killed many stands of trees within a few weeks. According to one observer, "The vast tracts of Douglas fir that stood green and venerable for generations [east of Yellowstone National Park] are peppered and painted with swaths of rusty red and gray. For Douglas fir, those are the colors of death" (Stark 2002, B-1).

Tens of millions of trees across the West were killed at a rate never seen before. Warmer temperatures accelerate the beetles' reproduction cycle, killing trees more quickly. Some types of beetles that propagated two generations a year were reproducing three times. "This is all due to temperature," said Barbara Bentz, a research entomologist with the U.S. Forest Service who is studying bark beetles. "Two or three degrees is enough to do it" (Wagner 2004). Outside Cody, Wyoming, an entire forest has been killed by the drought and beetles. "It used to be a nice spruce forest," said Kurt Allen, a Forest Service entomologist. "It's gone now. You're not going to get those conditions back for 200 or 300 years. We're really not going to have what a lot of people would consider a forest" (Wagner 2004).

Bill McEwen wrote in a letter to the *New York Times*: "I reside in the semi-arid West, where scientists are just beginning to understand the enormous synergistic impact of global warming, atmospheric drying (drought), and the explosion in insect populations that is killing many of our forests.... On a recent vacation to the Northwest, I drive through Sun Valley, Idaho. Around Sun Valley and the nearby Salmon River Valley, entire mountainsides of forest are now being destroyed by out-of-control bark beetle infestations" (McEwen 2004, A-16).

In Flagstaff, Arizona, home to the world's largest contiguous ponderosa pine forest, Tom Whitham wondered how much more devastation the drought and beetles would cause, and to what extent humans will contribute to it. "The thing that would make me really sad is if this was human caused," he said, glancing at the bare trees towering over his pickup truck. "If you lose a 200-year-old forest, you can't get it back" (Wagner 2004).

Deadly wildfires that burned at least 1,000 homes in southern California during late October 2003 were aggravated not only by fierce Santa Ana winds and 100-degree temperatures, drought, and very low humidity, but also by the deaths of more than a million mature pine trees killed during the previous year by bark beetle infestation.

According to a Canadian government projection, pine beetle infestation presents a "worst-case scenario" that could afflict 80 percent of British Columbia pine forests by the year 2020. "Mountain pine beetle outbreaks are stopped by severe winter weather or depletion of the host. The vast spatial extent of the outbreak implies that a weather-stopping event is unlikely," said the report (Baron 2004, A-3). By 2004, the beetle infestation already had spread through a broad swath of the British Columbia interior, especially around the Prince George area, extending south to the U.S. border.

INSECTS FIND NEW RANGES
IN MID-LATITUDE CITIES

Warming climate has favored the spread of many insects that previously were kept under control by seasonal frosts and freezes. Spreading insect infestations have been notable in many mid-latitude urban areas, among them London, England, and Boston, Massachusetts. According to James Meek, writing in the *London Guardian*,

> They're chomping in Chelsea, dining *al fresco* in Fulham and more than pecking at their food in Pimlico. Despite their fancy taste in London addresses they are neither posh nor particularly fussy. They are alien vine weevils, and they want to eat your plants. Two species of vine weevil previously unable to survive Britain's cold winters have been discovered in southwest London. One of the species has also been detected in Surrey and Cardiff and as far north as Edinburgh. Two new species, *otiorhynchus armadillo* and *otiorhynchus salicicola*, not previously known north of Switzerland, will add to gardeners' woes. (Meek [October 8] 2002, 6)

"This is probably the most serious new garden pest in recent memory," said Max Barclay, curator of beetles at the Natural History Museum in London, who has been following waves of insect infestations novel to Britain. Classed as beetles, the immigrating species are now prevalent in Chelsea, Victoria, Pimlico, and Fulham, causing significant damage to gardens, in some cases defoliating garden plants almost entirely. "It's very likely these weevils have been introduced to Britain through imported ornamental plants from Italy," said Barclay. "It looks like they're here to stay" (Meek [October 8] 2002, 6). Earlier springs and later, milder winters have drawn other new pests to Britain. "Of course, the fact [that] Mediterranean species are doing well in

Britain invites speculation about climate amelioration," said Barclay (p. 6).

After one of Boston's mildest winters on record in 2001–2002, residents and exterminators reported "one of the largest explosions of urban wildlife in recent memory" (Wilmsen 2002, B-1). Rats, ants, wasps, snakes, squirrels, rats, skunks, and other animals used the mild winter to procreate at times when the rigors of a New England winter usually would have killed many and kept the rest sexually dormant. Many small mammals had litters early and probably had one or more extra litters of young during the warm months of early 2002. (Many of the same creatures experienced the traditional perils of a New England winter during 2002–2003 and 2003–2004, however, when snow and cold returned with a vengeance.)

"I expect sharp increases in rat populations all over New England," said Bruce Colvin, a Lynnfield, Massachusetts, ecologist who is an expert on rats. "When you have a mild winter, there's more food and warm places for them to live, which means less stress. Instead of killing each other in the competition for food, they're all getting along and reproducing." Rat pups born during the winter were already mature enough by early spring to have litters of their own (Wilmsen 2002, B-1). Larger-than-usual numbers of carpenter ants and other insects, including bees and other wasps, were emerging in Boston by late March 2002, well ahead of their usual seasonal introduction in June.

By the first week of June 2002, also in Boston, pollen counts were at levels not usually observed until the end of the month. Epstein said that if carbon dioxide levels double, ragweed pollen counts will rise 61 percent. He anticipates that level will be reached by 2050. (Epstein wrote in the *Annals of Allergy, Asthma, and Immunology*, as reported in the *Boston Globe* [S. Smith 2002, A-1].) The record mild weather of the 2001–2002 winter was followed in New England by two severe winters that played a role in reducing pest populations to more usual levels.

Great Britain's rat population rose by nearly a third between 1998 and 2002. According to a report described in the *London Daily Mail*, by 2002 an estimated 60 million brown rats inhabited Great Britain's buildings, streets, sewers, and waterways. The newspaper reported that "the rodents narrowly outnumber the human population" (Utton 2003). Even Buckingham Palace called in exterminators after a rat infestation in its kitchens during 2002. The 29 percent, four-year rise in rat populations was blamed on milder, wetter winters as well as fast-food litter on which the rats feast. The amount of rubbish on London streets had grown by an estimated 80 percent in thirty-five years, according to the report. Rats breed quickly, reaching sexual maturity in eight weeks. They can have sex twenty times a day and can give birth every four weeks. A single pair of rats can produce as many as 2,000 offspring a year (Utton 2003).

Barrie Sheard, of the British National Pest Technicians' Association, which conducted the study, pointed to a rapid increase in rat numbers during the summer months, with warming a probable factor. Other factors also were at work. For example, many sewer systems built during the late nineteenth century have decayed, allowing more rats to reach the surface. To save money, many British town councils have started billing private property owners for rat control, which has caused many individuals to avoid reporting infestations. "Until these last four to five years, our summer months were always recognized as a period of the year when rat complaints were always at a far-reduced level," Sheard said. Peter Gibson, speaking for Keep Britain Tidy, stated, "The worrying thing is not just their numbers, but their behavior. Rats are not just in the sewers any more. They are coming out into the open, on to the streets, because they know that's where the most food is" (Utton 2003).

19 EFFECTS ON HUMAN HEALTH

When climate change scientists and diplomats met in Buenos Aires during 1998, they were greeted by news that mosquitoes carrying dengue fever had invaded more than a third of the homes in Argentina's most populous province, home to 14 million people. The *Aedes aegypti* mosquito appeared in Argentina in 1986; within twelve years, it was found in 36 percent of homes in Buenos Aires province, according to Dr. Alfredo Seijo of the Hospital Munoz. "*Aedes aegypti* now exists from the south of the United States to Buenos Aires province and this is obviously due to climatic changes which have taken place in Latin America over the past few years," Seijo told a news conference organized by the World Wildlife Fund at the U.N. climate talks in Argentina (Webb 1998).

Dengue fever, for which no vaccine exists, had nearly disappeared from the Americas by the 1970s. During the 1980s, however, the disease increased dramatically in South America, infecting over 300,000 people there by 1995. Also during 1995, Peru and the Amazon valley were especially hard hit by the area's largest epidemic of yellow fever since 1950, which is carried by the same mosquito that transmits dengue fever. The annual world incidence of dengue fever, which averaged about 100,000 cases between 1981 and 1985, averaged 450,000 cases a year between 1986 and 1990 (Gelbspan 1997, 149).

Dengue fever is one of a number of mosquito-vector diseases that have been increasing their coverage in many areas of the

Earth—climbing elevation in the tropics and rising in latitude in temperate zones—as global temperatures have warmed during the last quarter of the twentieth century. Rising temperatures and humidity increase the range of many illnesses spread by insects, including mosquitoes, warm-weather insects that die at temperatures below a range of 50–61 degrees F, depending on species. Dengue, a common disease in tropical regions, is a prolonged, flu-like viral infection that can cause internal bleeding, fever, and sometimes death. Dengue, which is sometimes called "breakbone fever," can be accompanied by headache, rash, and severe joint pain. The World Health Organization lists dengue fever as the tenth deadliest disease worldwide.

During 1995, an explosion of termites, mosquitoes, and cockroaches hit New Orleans, following an unprecedented five years without a killing frost. "Termites are everywhere. The city is totally, completely, inundated with them," said Ed Bordees, a New Orleans health official, who added that "the number of mosquitoes laying eggs has increased tenfold" (Gelbspan 1997, 15). The situation in New Orleans was aggravated not only by unusual warmth but also by above-average rainfall, totaling about eighty inches the previous year. Some of the 200-year-old oaks along New Orleans' St. Charles Avenue were found to have been eaten alive from the inside by billions of tiny, blind, Formosan termites.

The same year, dengue fever spread from Mexico across the border into Texas for the first time since records have been kept. At the same time, Colombia was experiencing plagues of mosquitoes and outbreaks of the diseases they carry, including dengue fever and encephalitis, triggered by a record heat wave followed by heavy rains.

Mild winters with a lack of freezing conditions allow many disease-carrying insects to expand their ranges. "Indeed," commented Paul Epstein of Harvard Medical School's Center for Health and the Global Environment, "fossil records indicate that when changes in climate occur, insects shift their range far more rapidly than do grasses, shrubs, and forests, and move to more favorable latitudes and elevations hundreds of years before larger animals do. 'Beetles,' concluded one climatologist, 'are better paleo-thermometers than bears' " (Epstein 1998).

WARMING AND EXCESS DEATHS WORLDWIDE

At least 150,000 people die needlessly each year as a direct result of global warming, three major U.N. organizations asserted late in 2003. Warming already has been a factor in a noticeable increase in malnutrition as well as outbreaks of diarrhea and malaria, the three largest killers in the poorest countries of the world, according to these reports. Diarmid Campbell-Lendrum, a World Health Organization scientist, said that estimates of deaths were extremely conservative. Furthermore, he said, the number of deaths attributable to rapid warming is expected to double during the coming thirty years. "People may say that this is a small total compared with the totals who die anyway, but these are needless deaths. We must do our best to take preventative measures," he said (Brown 2003, 19).

The report produced by the World Health Organization, the U.N. Environment Programme, and the World Meteorological Programme detailed how increased warmth has intensified the spread of diseases. The bacteria that cause diarrhea, spread mostly through unclean water and food, develop and diffuse more quickly in warmer temperatures and humidity. Dirty water is the largest killer of children less than five years of age. In Lima, Peru, a six-year study at a clinic set up to treat diarrheal complaints showed a 12 percent increase in cases for every 1 degree C rise in temperature during cooler months, and a 4 percent increase in warmer months (Brown 2003, 19). Similar results were found in a survey of eighteen Pacific islands. The problem is made worse by high rainfall or drought, during which water supplies become contaminated.

Many diseases spread by rats and insects also become more common in warmer weather. Malaria, dengue fever, and Lyme disease are all on the increase. Many threats can be curtailed by dispensing preventive medicine and providing clean water and sanitation. Climate change makes these issues more urgent, the U.N. report said. The combined effects of increased warmth and the greater volume of standing water brought by storms create malaria epidemics by providing breeding sites and an accelerated life cycle for mosquitoes. In Africa, where the death toll from malaria is highest, mosquitoes carrying the disease are spreading into mountain areas previously too cool for them to thrive, according to the report.

TEMPERATURE, HUMIDITY, AIR POLLUTION, AND HUMAN MALADIES

Aside from insect-vector diseases, warmer, more humid weather may aggravate urban air pollution (especially tropospheric ozone) that is a factor in many human maladies, such as asthma. Pim Martens of the Netherlands' Center for Integrative Studies (and a senior scientist at the University of Maastricht) pointed to a number of studies indicating that air pollution's effects on human health increase as temperatures rise: "Simultaneous exposure to heat and pollution appears to be more harmful than the sum of the individual effects" (Martens 1999, 535). Allergenic pollens and spores are more readily dispersed during hot, dry summers, Martens said.

Asthma is aggravated by heat (which increases the pollen production of many plants) as well as air pollution. According to the American Lung Association, more than 5,600 people died of asthma in the United States during 1995, a 45.3 percent increase in mortality over ten years and a 75 percent increase since 1980. Roughly a third of those cases occurred in children less than eighteen years of age. Since 1980, children under age five have experienced a 160 percent increase in asthma (Sierra Club 1999).

Poor and minority children are likely to develop asthma at worsening rates because of global warming combined with air pollution. As climate warms, allergens such as pollen and mold increase, interacting with urban pollutants such as ozone and soot to fuel a growing epidemic of asthma. "It is affecting the trees, the molds, the subsurface organisms," said Epstein. Reports indicate that asthma among U.S. preschool children aged three to five rose by 160 percent between 1980 and 1994. "This is a real wake-up call for people who think global warming is only going to be a problem way off in the future," said Christine Rogers, senior research scientist at the Harvard School of

Public Health. Rogers, Epstein, and the American Public Health Association worked together to report on these findings ("Findings" 2004).

"What happens to asthmatics in the heat and humidity? Well, we can't breathe," commented Barbara Mann, an author, teacher, and asthmatic who lives in Toledo, Ohio. "I'm permanently on three different prescription inhalers that I must use at regular intervals, four times a day—and that's just when I'm well. When I'm sick, there are antibiotics to bust up the hardened mucus in my lungs, along with other regimens of pills to aid the process. When my lungs are irritated by pollutants or by natural 'triggers' such as pollens or infected, my air passages swell shut, suffocating me" (Barbara Alice Mann, personal communication, August 3, 1999).

A study by the Sierra Club indicated that air pollution enhanced by global warming could be responsible for a number of human health problems, including respiratory diseases such as asthma, bronchitis, and pneumonia. According to Dr. Joel Schwartz, an epidemiologist at Harvard University, present-day air pollution concentrations are responsible for 70,000 early deaths per year and more than 100,000 excess hospitalizations for heart and lung disease in the United States. This could increase 10–20 percent in the United States as a result of global warming, with significantly greater increases in countries that are more polluted to begin with, according to Schwartz (Sierra Club 1999).

In addition to aggravation of specific diseases, persistent heat and humidity has a general debilitating influence on most warm-blooded animals, including humans. Farm animals especially are adversely affected when the air temperature remains higher than usual throughout the night. During the latter half of the twentieth century, according to Epstein, nighttime minimum temperatures over land areas rose at a rate of 1.86 degrees C per 100 years, and maximum temperatures rose at a 0.88 degrees C rate per 100 years during the same period (Epstein 1998).

GREENHOUSE GAS REDUCTION AND HEALTH

Luis Cifuentes and colleagues observed in *Science*: "The same actions that can reduce the long-term buildup of greenhouse gases—reductions in burning of fossil fuels—can also yield powerful, immediate benefits to public health by reducing the adverse effects of local air pollution." Their study cited estimates that 18,700 premature deaths per year could be avoided in the United States by reducing emissions from older, coal-fired electrical generating plants (Cifuentes et al. 2001, 1257). Deaths due to bronchial problems, heart disease, and other ailments would be reduced substantially. Another of the surveyed studies indicated that air pollution from traffic causes more deaths than traffic accidents.

The World Health Organization estimated that, in 1995, 460,000 avoidable deaths occurred annually worldwide "as a result of suspended particulate matter, largely from outdoor urban exposures" (Cifuentes et al. 2001, 1257). The authors noted that several types of pollution rise with temperatures. According to estimates developed by Cifuentes and colleagues, reducing fossil fuel air pollution in four of the world's largest cities (New York, Mexico City, São Paulo, and Santiago, Chile) could prevent 64,000 premature deaths between 1980 and 2000, reducing rates of infant mortality, asthma, cardiovascular problems, and respiratory ailments.

"The benefits of lowering emissions are immediate" because many of the gases emitted when fuels are burned are also pollutants, said George Thurston, one of the review's authors and an associate professor of environmental medicine at the New York University School of Medicine. "Universal studies have shown when air pollution levels go up, you get an increase in the numbers of deaths and hospital admissions, missed days at work and school, and other adverse effects," Thurston said (Surendran 2001, A-20).

Another study reported that alternative transportation policies initiated during the busy 1996 Summer Olympics in Atlanta "not only reduced vehicle exhaust and air pollutants such as ozone by about 30 per cent, they also decreased the number of acute asthma attacks by 40 per cent and pediatric emergency admissions by about 19 per cent (Surendran 2001, A-20).

GLOBAL WARMING AND DISEASE VECTORS

John T. Houghton, author of *Global Warming: The Complete Briefing* (1997), believes that global warming will accelerate the spread of many diseases from the tropics to the middle latitudes. Malaria could increase from its present level, Houghton warns. "Other diseases which are likely to spread for the same reason are yellow fever, dengue fever, and . . . viral encephalitis," he wrote (p. 132). After 1980, small outbreaks of locally transmitted malaria occurred in Texas, Georgia, Florida, Michigan, New Jersey, New York, and California, usually during hot, wet spells. World-wide, according to Epstein, between 1.5 and 3 million die of malaria each year, mostly children. Mosquitoes and parasites that carry the disease have evolved immunities to many insecticides.

According to Epstein, "If tropical weather is expanding it means that tropical diseases will expand. We're seeing malaria in Houston, Texas" (Glick 1998). Epstein suggested that a resurgence of infectious diseases may be one result of global warming. Warming may appear beneficial at first, Epstein said. Initially, some plants benefit from additional warmth and moisture, an earlier spring, and more carbon dioxide and nitrogen in the air. "But," he cautioned, "warming and increased CO_2 can also stimulate microbes and their carriers" (Epstein 1998).

Since 1976, Epstein reported, thirty diseases have emerged that are new to medicine. Old ones, such as drug-resistant tuberculosis, have been given new life by new diseases (such as HIV/AIDS) that com-promise the human immune system. By 1998, tuberculosis was claiming 3 million lives annually around the world. "Malaria, dengue, yellow fever, cholera, and a number of rodent-borne viruses are also appearing with increased frequency," Epstein reported. During 1995, mortality from infectious diseases attributed to causes other than HIV/AIDS rose 22 percent above the levels of fifteen years earlier in the United States. Adding deaths complicated by HIV and AIDS, deaths from infectious diseases have risen 58 percent in fifteen years (Epstein 1998). The IPCC included a chapter on public health in an update of its 1990 assessment, which concluded: "Climate change is likely to have wide-ranging and mostly adverse impacts on human health, with significant loss of life" (Taubes 1997).

Andrew Haines asserted, in Jeremy Leggett's *Global Warming: The Greenpeace Report* (1990): "Although winter bronchitis and pneumonia may be reduced [by global warming], it is quite likely that hay fever and perhaps asthma could increase. A combination of increase in temperature with increasing levels of tropospheric ozone could have clinically important effects, particularly in patients with asthma and chronic obstructive airways disease" (Haines 1990, 154).

Epstein (1998) identified three tendencies in global climate change and related each to an increasingly virile environment for infectious diseases. The three indicators are:

1. Increased air temperatures at altitudes of two to four miles above the surface in the Southern Hemisphere.
2. A disproportionate rise in minimum temperatures, in either daily or seasonally averaged readings.
3. An increase in extreme weather events, such as droughts and sudden heavy rains.

"There is growing evidence for all three of these tell-tale 'fingerprints' of enhanced greenhouse warming," he stated.

A Sierra Club (1999) study indicated that a lengthy El Niño event during the middle 1990s provided an indication of how sensitive some diseases can be to changes in climate. The study cited evidence that warming waters in the Pacific Ocean contributed to a severe outbreak of cholera that led to thousands of deaths in Latin American countries during the 1990s. According to health experts quoted by the Sierra Club study, "The current outbreak [of dengue fever], with its proximity to Texas, is at least a reminder of the risks that a warming climate might pose." The study concluded: "While it is difficult to prove that any particular outbreak was caused or exacerbated by global warming, such incidents provide a hint of what might occur as global warming escalates."

Willem Martens and colleagues, writing in *Climatic Change*, attempted to sketch how a warmer, wetter climate would affect transmission of three vector-borne diseases: malaria, schistosomiasis, and dengue fever. The scholars anticipated that the periphery of the currently endemic areas will expand with global warming, with diseases notable at higher elevations in the tropics, an expectation that has been borne out by

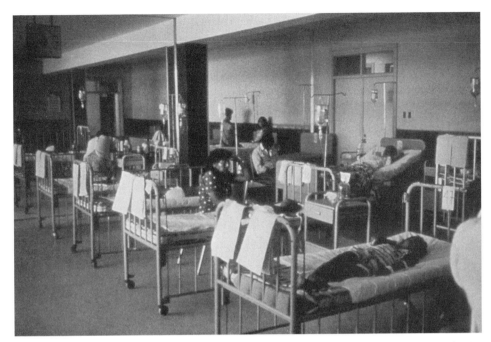

A Peru hospital waiting room that was converted to an emergency cholera ward during an epidemic. Courtesy of the Centers for Disease Control and Prevention.

several observers. Martens and colleagues expected that "the increase in epidemic potential of malaria and dengue transmission may be estimated at 12 to 27 per cent and 31 to 47 per cent respectively" (Martens, Jetten, and Focks 1997, 145). In contrast, they forecast that the transmission potential of schistosomiasis may decrease 11–17 percent.

The incidence of eastern equine encephalitis, which attacks both horses and humans, has been increasing in parts of the United States, although transmission to humans is still considered rare. Prince Georges County, adjacent to Washington, D.C., reported five cases in 1996, with the origins of contraction unknown (Bloomfield and Showell 1997). Early symptoms include fever, headache, drowsiness, and muscle pain, followed by disorientation, weakness, seizures, and coma. Sixty percent of cases are fatal, and most survivors suffer permanent neurological damage. Mild winters and wet springs, expected to become more likely with global warming, are associated with increased risk of eastern equine encephalitis.

Another disease that may become more common in the temperate zones of a warmer world is diarrhea, which during the 1990s was killing more than 3 million children a year worldwide, mainly in the tropics of Asia, Africa, and the Americas. The bacteria that cause diarrhea thrive in warm weather, especially after heavy rainfall. Warm, moist weather also promotes the growth and activity of flies and cockroaches.

The effects of warming-induced diseases extend to the oceans. Epstein commented, "Warming—when sufficient nutrients are present—may also be contributing to the proliferation of coastal algal blooms. Harmful algal blooms of increasing extent, duration, and intensity—and involving novel, toxic species—have been reported around the world since the 1970s. Indeed, some scientists feel that the worldwide increase in coastal algal blooms may be one of the first biological signs of global environmental change" (Epstein 1998).

MALARIA IN A WARMER WORLD

In the tropics, elevation has long been used to shield human populations from diseases that are endemic in the lowlands. With global warming, mosquito-borne diseases have been reaching higher altitudes, affecting people with little or no immunity. According to Pim Martens, "A minor temperature rise will be sufficient to turn the populated African highlands into an area that is suitable for the malaria mosquito and parasite" (Martens 1999, 537).

During 1997, malaria ravaged large areas of Papua New Guinea at an elevation of 2,100 meters, notably higher than the 1,200–2,000 meters that heretofore had provided a barrier to the disease in different parts of central and southern Africa. In northwestern Pakistan, according to Martens, a rise of about half a degree C in the mean temperature was a factor in a rising incidence of malaria there, from a few hundred cases a year in the early 1980s to 25,000 in 1990 (Martens 1999, 537). Whereas most strains of malaria can be controlled, drug-resistant strains were proliferating late in the twentieth century.

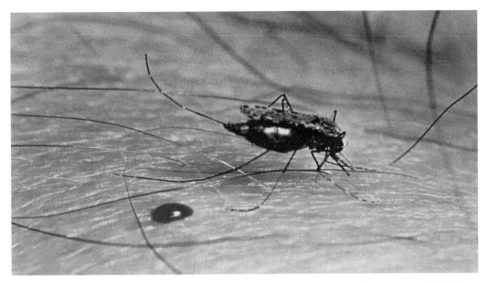

An *Anopheles gambiae* mosquito, a vector for malarial parasites. Courtesy of the Centers for Disease Control.

Writing in the *Bulletin of the American Meteorological Society* (March 1998), Epstein and seven co-authors described the spread of malaria and dengue fever to higher altitudes in tropical areas of the Earth because of warmer temperatures. Rising winter temperatures also have allowed disease-bearing insects to survive in areas previously closed to them. According to Epstein, frequent flooding, which is associated with warmer temperatures, also promotes the growth of fungus and provides excellent breeding grounds for large numbers of mosquitoes. The flooding caused by Hurricane Floyd and other storms in North Carolina during 1999 is cited by some as a real-world example of how global warming promotes conditions ideal for the spread of diseases imported from the tropics (Epstein et al. 1998).

According to the IPCC's projections for human health, a rise in global average temperatures of 3–5 degrees C by 2100 could lead to 50–80 million additional cases of malaria a year worldwide, "primarily in tropical, subtropical and less well-protected temperate-zone populations" (Watson, 2001). Italy experienced a brief outbreak of malaria during 1997. Hadley Climate Center researchers expect the same disease to reach the Baltic States by 2050. In parts of the world where malaria is now unknown, most people have no immunity (Brown 1999). The World Health Organization projected that warmer weather will cause tens of millions additional cases of malaria and other infectious diseases. The Dutch health ministry anticipates that more than a million people may die annually as a result of the impact of global warming on malaria transmission in North America and northern Europe (Epstein 1999, 67).

Malaria could return to Britain as an endemic disease, scientists at the University of Durham warned as they announced a plan to produce a "risk map" showing which areas were most likely to suffer an outbreak (Connor 2001, 14). With millions of tourists visiting malaria-infested regions of the world, the risk of the disease making a comeback is further increased by global warming, which expands mosquito habitat in the United Kingdom, said Rob Hutchinson, an entomologist at the university, at the annual meeting of the Royal Entomological Society in Aberdeen. He said that, of the 25 million overseas visitors who came to Britain in 1999, about 260,000 came from Turkey and the countries of the former Soviet Union, where vivax malaria was endemic and health care was poor (p. 14).

By the late 1990s, malaria had been transmitted by mosquitoes as far north as Toronto, Canada, according to Epstein. "The extreme events we are seeing today in Nicaragua and Honduras [as a result of Hurricane Mitch in 1998] are spawning outbreaks of cholera and dengue fever with new breeding sites for mosquitoes and increased water-borne diseases," Epstein said (Webb 1998). In northerly latitudes, nighttime and winter temperatures have warmed twice as fast as overall global temperatures since 1950, meaning that fewer pests are being killed by frost in the southern reaches of the temperate zones. Humidity also has increased in many regions, including much of the eastern United States, helping mosquitoes breed. Disease-carrying mosquitoes usually require a certain level of temperature *and* humidity to survive.

In May 1995, researchers in the Netherlands and in England estimated the increase in malaria's geographic range that could occur if the

IPCC's projections for global warming prove correct. These researchers concluded that, in tropical regions, the epidemic potential of the mosquito population would double. In temperate climates, according to these projections, the epidemic potential could increase a hundredfold. Furthermore, "there is a real risk of reintroducing malaria into non-malarial areas, including parts of Australia, the United States, and southern Europe" (Environmental Research Foundation 1995).

By the late 1970s, dwindling investments in public health programs, growing resistance to insecticides among some species, and prevalent environmental changes (such as deforestation) contributed to a widespread resurgence of malaria, according to Epstein. By the late 1980s, he reports, large epidemics of malaria were being associated with warm, wet weather. Between 1993 and 1998 worldwide incidence of malaria quadrupled. "Malaria is now found in higher-elevations in central Africa and could threaten cities such as Nairobi, Kenya (at about 5,000 feet, roughly the elevation of Denver, Colorado), as freezing levels have shifted higher in the mountains. In the summer of 1997, for example, malaria took the lives of hundreds of people in the Kenyan highlands, where populations had previously been unexposed" (Epstein 1998).

Between 1970 and 1995, the lowest level at which freezing occurs has climbed about 160 meters higher in mountain ranges, from 30 degrees north to 30 degrees south latitude, based on radiosonde data analyzed at NOAA's Environmental Research Laboratory. This shift corresponds to a warming at these elevations of about 1 degree C (almost 2 degrees F), which is nearly twice the average warming that had been documented over the Earth as a whole by the end of the twentieth century (Epstein 1998). As higher elevations warm, mosquito-vector diseases are ascending tropical mountainsides around the world. Bill Weinburg, in *Native Americas*, describes changes provoked by warming in the mountains of Mexico:

Dr. Juan Blechen Nieto, a Cuernavaca physician, traveled through the Sierra del Sur on a survey of local health conditions in November 1998, and found an alarming incidence of dengue fever and malaria. "These are diseases that are traditionally associated with lowland coastal regions, and are now appearing in the Sierra del Sur," he told me.... Indians in highland Oaxaca communities tell me they have mosquitoes now, for the first time. This has to do with deforestation impacting local and regional climate. It gets hotter, and the undergrowth that comes up after forests are destroyed provides a habitat for pests. (Weinburg 1999, 58–59)

MORTALITY FROM HEAT WAVES

Historically, heat stress has been the foremost weather-related cause of death in the United States. During the second half of the twentieth century, however, even as temperatures rose, the rate of heat-related deaths declined dramatically due to increased use of air conditioning, better medical care, and increased public awareness of heat stress effects. Robert Davis, associate professor of environmental sciences at the University of Virginia, and colleagues studied heat-related mortality in twenty-eight major U.S. cities from 1964 through 1998. He found that the heat-related death rate, 41 per million people a year in the 1960s and 1970s, declined to 10.4 per million during the 1990s (Davis et al. 2003).

Cities in the United States have ten more hot nights a year than forty years ago, Cornell University climate researchers have found. Although summers are heating up in urban areas, in rural areas temperatures have remained more constant, said Arthur DeGaetano, associate professor of Earth and atmospheric sciences at Cornell. "What surprised me was the difference in the extreme temperature trends between rural and urban areas," said DeGaetano. "I expected maybe a 25 percent increase for the urban areas compared to the rural ones. I didn't expect a 300 percent increase across the United States" ("Hot Times" 2002).

Rural areas experienced an average increase of only three warm nights a year in the same period, according to this study. "This means that cities and the suburbs may be contributing greatly to their own heat problems," DeGaetano said. "Greenhouse gases could be a factor, but not the one and only cause. There is natural climate variability, and you tend to see higher temperatures during periods of drought" ("Hot Times" 2002).

DeGaetano and colleagues classified a warm night as a minimum of 70 degrees F in the eastern, southern, and midwestern United States. In the Southwest, a low of 80 degrees was considered a warm night. Since the beginning of the twentieth century, almost three-quarters of the climate-reporting stations examined in the study have shown an increase in the number of very warm nights, according to the study ("Hot Times" 2002).

Laurence S. Kalkstein has estimated that a doubling of the carbon dioxide level in the atmosphere could increase heat-related mortality to seven times present levels if acclimatization is not factored in. With acclimatization (human adaptation to higher temperatures), the estimated increase in heat-wave mortality estimated by Kalkstein is four times the present rate (Kalkstein 1993, 1397). Kalkstein observed that each urban area has its own "temperature threshold" at which the death rate from heat prostration rises rapidly. Seattle, for example, has a lower threshold than Dallas. "Mortality rates in warmer cities seemed to be less affected no matter how high the temperature rose," Kalkstein wrote (p. 1398). He suggested that residents of urban areas in poor countries will find adaptation more difficult because of limited access to air conditioning.

A region need not be impoverished to suffer a stunning degree of mortality from heat. One analysis put Europe's death toll at 35,000 or more during the scorching summer of 2003, which is described in Chapter 6: "Watching the Thermometer." The one common thread in most of these deaths was lack of access to air conditioning, which had not heretofore been considered necessary in much of Europe.

A 1988 Environmental Protection Agency (EPA) study estimated that heat-wave mortality in fifteen large U.S. cities could rise from 1,200 a year to 7,500 a year if carbon dioxide levels double (Schneider 1989, 182). Another EPA report indicated in the late 1980s that a 4 degree C rise in San Francisco's temperature would raise ozone levels there by about 20 percent, with attendant health effects (p. 183). According to Anthony J. McMichael, professor of epidemiology at the London School of Hygiene and Tropical Medicine, higher summer temperatures in both temperate and tropical regions could increase the rates of serious illness and death from heat-related causes by as much as six times the current level, with the greatest impact falling on the sick and elderly (McMichael 1993, 143).

Cities tend to emit and absorb heat more quickly than surrounding countryside due to a number of reasons having little to do with the basic atmospheric physics of global warming. The larger a city and the more dense its degree of urbanization, the greater the warming. The "urban heat-island effect" was first identified by a meteorologist, Luke Howard, in 1818. Extra heat is produced in urban areas by a city's many sources of waste heat—from building heating and air conditioning as well as from motor vehicles, among other sources. Heating also increases when open fields and forests become streets, sidewalks, parking lots, and buildings. The dark colors of city structures, especially asphalt streets and parking lots that make up as much as 30 percent of many urban surfaces, have a very low albedo (reflectivity), so most of the sun's heat energy is absorbed, not reflected.

Cities also warm more rapidly than surrounding countryside because they are usually drier and have less surface water and plant mass (both of which cool the air through evaporation) than most rural areas. Furthermore, as new housing and businesses spread from urban areas, some of the cities' urban heat follows them, spreading in widening suburban circles. In Japan, suburban areas near Tokyo have experienced temperature rises of between 2 and 3 degrees C in ten years, following urbanization. In a compact urban area such as Manhattan Island, the total heat generation of the city can add quite substantially to solar radiation. By one estimate, the heat energy generated by motor vehicles and space heating on Manhattan during an average winter day sometimes exceeds that of incoming solar radiation (Weiner 1990, 262).

Robert Balling, who professes skepticism regarding greenhouse warming as a global issue, is something of an expert on the urban heat-island effect in his hometown, Phoenix, Arizona. Since World War II, when many Phoenix residents slept on porches outdoors, average summertime lows in Phoenix have risen above the human comfort zone. Average summertime lows have risen from 73 degrees to more than 80 degrees F during the last half of the twentieth century. During the same fifty years, the Phoenix area's human population has increased by nearly twenty times, from roughly 150,000 to 2.8 million. Average daytime high temperatures during summer in Phoenix have remained roughly the same during the same half-century, at between 102 and 104 degrees F. Dale Quattrochi, senior research scientist at NASA's Global Hydrology and Climate Center in Huntsville, Alabama, estimated that Phoenix temperatures likely will increase as much as 15 to 20 degrees F over historic averages during the next several decades (Yozwiak 1998).

WARMING AND WEST NILE VIRUS

Until 1999, West Nile virus had never even been detected in North America. No one knows exactly how the virus reached the United States. Once it arrived, however, West Nile virus spread rapidly across the continent; by 2001 it had infected twenty-nine species of mosquitoes, 100 species of birds, and many mammals, including humans. By the summer of 2002, West Nile had reached thirty-six states, as well as the southern regions of eastern Canada (Grady 2002, F-2). By 2003, most of the continental United States was reporting West Nile virus. Global warming may be a contributing factor, due to warm winters and pervasive summer droughts that seem to favor the spread of the mosquito-borne virus.

The disease initially brings fever, aches, and profound fatigue, sometimes followed by paralysis and other neurological complications, including meningitis and encephalitis, which can leave a victim physically disabled and brain damaged. Between 1999 and 2004, West Nile virus infected more than 16,000 people in the United States, killing more than 600 and afflicting 6,500 others with severe neurological problems (Chase 2004, A-1).

Regarding West Nile virus, Epstein said that drought helps the mosquito species *Culex pipiens*, which plays a major role in spreading the disease. Epstein added that drought also may wipe out darning needles, dragonflies, and amphibians, which destroy mosquitoes. Drought also may aid the spread of infection by drawing thirsty birds to the pools and

Centers for Disease Control director Julie Louise Gerberding, speaking at a West Nile Virus press briefing, August 2003. Courtesy of the Centers for Disease Control.

puddles where mosquitoes breed. "Hot weather plays a role, too," Epstein said. "Warmth increases the rate at which pathogens mature inside mosquitoes" (Grady 2002, F-2).

HAY FEVER ARRIVES EARLY AND STAYS LATE

English scientists (as well as street-level observers) have maintained that the allergy season is arriving earlier because of global warming. Many trees and grasses are flowering sooner and for extended periods, creating more of the pollen that is the main trigger of hay fever. In 2002, the hay fever season began as early as January 30, according to research by Tim Sparks of the Centre for Ecology and Hydrology. The same research also indicated that varieties of common grass were flowering up to thirteen days earlier than in 2001. Hazel and birch trees also were causing allergic reactions in many hay fever sufferers. By early February 2003, stated Sparks, hazel trees were in flower (Chapman 2003, 23).

These findings were part of the world's largest phenological survey by the Centre for Ecology and Hydrology and the Woodland Trust. Jean Emberlin of the British National Pollen Research Unit said: "Last year, the grass pollen season was exceptionally long because the weather was wet and warm. The season extended into August instead of ending in July." Hay fever, otherwise known as seasonal allergic rhinitis, is caused by an allergy to pollens and fungal spores. Young people appear to be worst hit, with 36 percent suffering from it. The figure in the wider population is 15–25 percent, with rates doubling since 1965 (Chapman 2003, 23).

In the past, London hay fever sufferers usually have not stocked up on remedies in earnest until the beginning of May. Pharmacies in 2003 reported a 30 percent increase in sales each week between mid-March and mid-April. Muriel Simmonds, chief executive of Medical-Charity Allergy U.K., cited "definite evidence" that hay fever was returning to the British Isles earlier year after year. Simmonds' organization has moved National Allergy Week from June to May because of the earlier onset of hay fever. "This is the earliest I've ever seen it in London," she said (Galloway and Rhodes 2003, 16).

The epidemic seemed to be worldwide. David Adam, writing in the *London Guardian*, summarized the world hay fever situation:

They are sneezing in Stockholm, throats are itchy in India and Irish eyes are streaming. From Algeria to Iceland and Hong Kong to Aberdeen, record numbers of people are suffering the misery of hay fever—and it's getting worse. Everything from global warming to air pollution is conspiring to make hay fever the number-one global irritant. The figures are truly remarkable. The number of people rubbing their eyes in British doctor's surgeries [offices] has risen fivefold since the 1950s, and about a quarter of people in the United Kingdom are now believed to be hay-fever sufferers. (Adam 2003, 4)

Similar trends have been seen across Europe, Adam wrote. Cases of hay fever doubled or even trebled in Sweden and Finland during the 1970s and 1980s, whereas Swiss surveys show about one in ten are affected, an increase from fewer than one in 100 when a similar count was made in 1926. About 40 percent of the people in Australia and the United States said they suffered from hay fever or similar allergies by 2002. Areas such as West Africa, where hay fever was once all but unknown, have been reporting it as well (Adam 2003, 4).

HEALTH BENEFITS FROM WARMING?

Pim Martens has written that, although the overall impact of global warming on human health is expected to be markedly negative, humans may experience a few positive outcomes. Some diseases that thrive in cold weather (such as influenza) may find their ranges and effects reduced in a warmer world. The elderly might die less frequently of cardiovascular and pulmonary ailments, which peak during cold weather. "Whether the milder winters could offset the mortality during the summer heat waves is one of the questions that demands further research," Martens wrote (Martens 1999, 535).

Countering the views of Epstein and others, some health researchers contend that global warming will do little to increase the incidence of tropical diseases. "For mosquito-borne diseases such as dengue, yellow fever, and malaria, the assumption that warming will foster the spread of the vector is simplistic," contended Bob Zimmerman, an entomologist with the Pan American Health Organization (PAHO). Zimmerman pointed out that, in the Amazon basin, more than twenty species of *Anopheles* mosquitoes can transmit malaria, and each is adapted to a different habitat: "All of these are going to be impacted by rainfall, temperature, and humidity in different ways. There could actually be decreases in malaria in certain regions, depending on what happens." Virologist Barry Beaty of Colorado State University in Fort Collins, Colorado, agreed with Zimmerman: "You don't have to be a rocket scientist to say we've got a problem," he says. "But global warming is not the current problem. It is a collapse in public-health measures, an increase in drug resistance in parasites, and an increase in pesticide resistance in vector populations. Mosquitoes and parasites are efficiently exploiting these problems" (Taubes 1997).

Countering the majority view that a warmer world will spread malaria, David J. Rogers and Sarah E. Randolph, using their own models, wrote in *Science* that even extreme rises in temperature will not spread the disease. They argued that the spread of malaria is too poorly understood to forecast into the future based on temperature as a singular variable. For example, the "Dengie marshes" of Essex in England, a breeding ground for malaria-carrying mosquitoes in the seventeenth century, have dried up, making an increase in temperatures not a factor vis-à-vis malaria's transmission. Malaria is not a new disease in the temperate zones. It was common in the Roman Empire. A British invasion of Holland in 1806 failed to drive out French troops because large numbers of British became ill with malaria. Malaria was a public health problem in most of the eastern United States during warm, humid summers before medications were developed for it about a century ago.

Paul Reiter, a dengue expert with the Centers for Disease Control and Prevention's Puerto Rico office, argued against the relative importance of climate in human disease by pointing to periods in the past when malaria and other tropical diseases pervaded cooler regions. He argued that the spread of malaria is more closely linked to deforestation, agricultural practices, human migration, poor public health services, civil war, strife, and natural disasters. "Claims that malaria resurgence is due to climate change ignore these realities and disregard history," he wrote in an article about malaria's spread through England during the Little Ice Age, which began about 1450 and lasted for several hundred years during a period that was cooler than today (McFarling 2002, A-7).

S. I. Hay and colleagues investigated long-term meteorological trends in four high-elevation sites in East Africa, where increases in malaria have been reported during the past two decades. "Here we show that temperature, rainfall, vapor pressure, and the number of months suitable for *P. falciparum* transmission have not changed significantly during the past century or during the period of reported malaria resurgence." Therefore, they find that associations between resurgence of malaria and climate change at high altitudes in these areas "are overly simplistic" (Hay et al. 2002, 905).

REFERENCES:
PART V. FLORA
AND FAUNA

Adam, David. "Hatchoooooh! Record Numbers of People Are Complaining of Hay Fever." *London Guardian*, June 18, 2003, 4.

"Amazon Deforestation Causing Global Warming, Brazilian Government Says." British Broadcasting Corporation International reports, December 10, 2004. (Lexis).

Asner, Gregory P., David E. Knapp, Eben N. Broadbent, Paulo J. C. Oliveira, Michaael Keller, and Jose N. Silva. "Selective Logging in the Brazilian Amazon." *Science* 310 (October 21, 2005): 480–481.

Baron, Ethan. "Beetles Could Chew Up 80 Per Cent of B.C. Pine: Report: 'Worst-Case Scenario' by 2020 Blamed on Global Warming." *Ottawa Citizen*, September 12, 2004, A-3.

Benton, Michael J. *When Life Nearly Died: The Greatest Mass Extinction of All Time.* London: Thames and Hudson, 2003.

"Bitter Pill: The Northward Spread of the Okinawan Goya, Warm-Weather." Asahi News Service (Japan), January 29, 2003. (Lexis).

Bloomfield, Janine, and Sherry Showell. *Global Warming: Our Nation's Capital at Risk.* Environmental Defense Fund, 1997. www.edf.org/pubs/Reports/WashingtonGW/index.html.

Both, C., and M. E. Visser. "Adjustment to Climate Change Is Constrained by Arrival Date in a Long-Distance Migrant Bird." *Nature* 411 (May 17, 2001): 296–298.

Bradshaw, William E., and Christina M. Holzapfel. "Genetic Shift in Photoperiodic Response Correlated with Global Warming." *Proceedings of the National Academy of Sciences* 98 (25) (December 4, 2001): 14509–14515.

Brown, Paul. "Global Warming Kills 150,000 a Year: Disease and Malnutrition the Biggest Threats, United Nations Organisations Warn at Talks on Kyoto." *London Guardian*, December 12, 2003, 19.

———."Global Warming: Worse Than We Thought." *World Press Review*, February, 1999, 44.

Browne, Anthony. "How Climate Change Is Killing Off Rare Animals: Conservationists Warn That Nature's 'Crown Jewels' Are Facing Ruin." *London Observer*, February 10, 2002, 15.

———, and Paul Simons. "Euro-Spiders Invade as Temperature Creeps Up." *London Times*, December 24, 2002, 8.

"Carbon Pollution Wreaking Havoc with Amazonian Forest." Agence France Presse, March 10, 2004. (Lexis).

Chapman, James. "Early Spring Misery for 12 Million Hay Fever Sufferers." *London Daily Mail*, February 4, 2003, 23.

Chase, Marilyn. "As Virus Spreads, Views of West Nile Grow Even Darker." *Wall Street Journal*, October 14, 2004, A-1, A-10.

Choi, Charles. "Rainforests Might Speed Up Global Warming." United Press International, April 21, 2003. (Lexis).

Cifuentes, Luis, Victor H. Borja-Aburto, Nelson Gouveia, George Thurston, and Devra Lee Davis. "Hidden Health Benefits of Greenhouse Gas Mitigation." *Science* 252 (August 17, 2001): 1257–1259.

Clark, D. A., S. C. Piper, C. D. Keeling, and D. B. Clark. "Tropical Rain Forest Tree Growth and Atmospheric Carbon Dynamics Linked to Interannual Temperature Variation during 1984–2000." *Proceedings of the National Academy of Sciences* 100 (10) (May 13, 2003): 5852–5857.

Clarke, Tom. "Boiling Seas Linked to Mass Extinction: Methane Belches May Have Catastrophic Consequences." *Nature*, August 22, 2003. http://info. nature.com/cgi-bin24/DM/y/eLodoBfHSKoChoDYyoAL.

Collins, Simon. "Birds Starve in Warmer Seas." *New Zealand Herald*, November 14, 2002. (Lexis).

Connor, Steve. "Global Warming May Wipe Out a Fifth of Wild Flower Species, Study Warns." *London Independent*, June 17, 2003. (Lexis).

———. "Malaria Could Become Endemic Disease in U.K." *London Independent*, September 12, 2001, 14.

———. "World's Wildlife Shows Effects of Global Warming." *London Independent*, March 28, 2002, 11.

Cookson, Clive. "Global Warming Triggers Epidemics in Wildlife." *London Financial Times*, June 21, 2002, 4.

Davidson, E. A., and A. I. Hirsch. "Carbon Cycle: Fertile Forest Experiments." *Nature* 411 (May 24, 2001): 431–433.

Davis, Robert E., Paul C. Knappenberger, Patrick J. Michaels, and Wendy M. Novicoff. "Changing Heat-Related Mortality in the United States." *Environmental Health Perspectives*, July 23, 2003. doi:10.1289/ehp.6336. http://dx.doi.org.

"Deforestation Could Push Amazon Rainforest to Its End." UniScience News Net, July 3, 2001. www.Unisci.com.

Derbyshire, David. "Baffled Bumble Bee Lured Out Early by Changing Climate." *London Daily Telegraph*, March 12, 2004, 15.

———. "Global Warming Fails to Boost Butterfly Visitors." *London Daily Telegraph*, November 1, 2001, 13.

Eilperin, Juliet. "Warming Tied to Extinction of Frog Species." *Washington Post*, January 12, 2006, A-1. www.washingtonpost.com/wp-dyn/content/article/2006/01/11/AR2006011102121_pf.html.

Environmental Research Foundation. "Rachel's #466: Warming & Infectious Diseases." Annapolis, Maryland, November 2, 1995. www.igc.apc.org/awea/wew/othersources/rachel466.html.

Epstein, Paul R. "Climate, Ecology, and Human Health." December 18, 1998a. www.iitap.iastate.edu/gccourse/issues/health/health.html.

———. "Profound Consequences: Climate Disruption, Contagious Disease, and Public Health." *Native Americas* 16 (3/4) (Fall/Winter 1999): 64–67.

———, Henry F. Diaz, Scott Elias, Georg Grabherr, Nicohlas E. Graham, Willem J. M. Martenset, et al. "Biological and Physical Signs of Climate Change: Focus on Mosquito-borne Diseases." *Bulletin of the American Meteorological Society* 79 (3) (March 1998b): 409–417.

Fahy, Declan. "Nature Charts Its Own Change: Irish Researchers Are Finding Signs of Climate Change in Trees and Bird Species." *Irish Times*, September 13, 2001. (Lexis).

"Findings." *Washington Post*, April 30, 2004, A-30.

Fitter, A. H., and R.S.R. Fitter. "Rapid Changes in Flowering Time in British Plants." *Science* 296 (May 31, 2002): 1689–1691.

Fountain, Henry. "Observatory: Early Birds and Worms." *New York Times*, May 22, 2001, F-4.

———. "Observatory: Rice and Warm Weather." *New York Times*, June 29, 2004, F-1.

———. "Observatory: Threat to Rice Crops." *New York Times*, December 12, 2000, F-5.

Galloway, Elaine, and Chloe Rhodes. "Warm Spell Brings Early Start to Hay-Fever Misery." *London Evening Standard*, April 14, 2003, 16.

Gelbspan, Ross. *The Heat Is On: The High Stakes Battle over Earth's Threatened Climate*. Reading, MA: Addison-Wesley Publishing Co., 1997.

Gjerdrum, Carina, Anne M. J. Vallee, Colleen Cassady St. Clair, Douglas F. Bertram, John L. Ryder, and Gwylim S. Blackburn. "Tufted Puffin Reproduction Reveals Ocean Climate Variability." *Proceedings of the National Academy of Sciences* 100 (16) (August 5, 2003): 9377–9382.

Glick, Patricia. *Global Warming: The High Costs of Inaction*. San Francisco: Sierra Club, 1998. www.sierraclub.org/global-warming/inaction.html.

"Global Warming Is Changing Tropical Forests." Environment News Service, August 7, 2002. http://ens-news.com/ens/aug2002/2002-08-07-01.asp.

Grabherr, G., M. Gottfried, and H. Pauli. "Climate Effects on Mountain Plants." *Nature* 339 (1994): 448–451.

Grady, Denise. "Managing Planet Earth: On an Altered Planet, New Diseases Emerge as Old Ones Re-emerge." *New York Times*, August 20, 2002, F-2.

Gugliotta, Guy. "Warming May Threaten 37 Per Cent of Species by 2050." *Washington Post*, January 8, 2004, A-1. www.washingtonpost.com/wp-dyn/articles/A63153-2004Jan7.html.

Haines, Andrew. "The Implications for Health." In *Global Warming: The Greenpeace Report*, ed. Jeremy Leggett, 149–162. New York: Oxford University Press, 1990.

Hallam, Anthony, and Paul Wignall. *Mass Extinctions and Their Aftermath*. Oxford: Oxford University Press, 1997.

Hann, Judith. "Spring Wakes Early, but Will Autumn Lie in Late Again? What Will Tomorrow's World Look Like?" United Kingdom Woodland Trust, 2002. www.woodland-trust.org.uk/news/subindex.asp?aid=328.

Harvell, C. Drew, Charles E. Mitchell, Jessica R. Ward, Sonia Altizer, Andrew P. Dobson, Richard S. Ostfeld, et al. "Climate Warming and Disease Risks for Terrestrial and Marine Biota." *Science* 296 (June 21, 2002): 2158–2162.

Hay, S. I., J. Cox, D. J. Rogers, S. E. Randolph, D. I. Stern, G. D. Shanks, et al. "Climate Change and the Resurgence of Malaria in the East African Highlands." *Nature* 425 (February 21, 2002): 905–909.

Hogan, Treacy. "Still Raining in Costa del Ireland." *Belfast Telegram*, November 26, 2002. (Lexis).

"Hot Times in the City Getting Hotter." Environment News Service, September 27, 2002. http://ens-news.com/ens/sep2002/2002-09-27-09.asp#anchor8.

Houghton, John. *Global Warming: The Complete Briefing*. Cambridge, U.K.: Cambridge University Press, 1997.

Hungate, Bruce A., Peter D. Stiling, Paul Dijkstra, Dale W. Johnson, Michael E. Ketterer, Graham J. Hymus, et al. "CO_2 Elicits Long-Term Decline in Nitrogen Fixation." *Science* 304 (May 28, 2004): 1291.

Ingham, John. "Stingers Thrive as the Country Gets Warmer: Invasion of the Scorpions." *London Express*, June 18, 2004, 40.

Inkley, D. B., M. G. Anderson, A. R. Blaustein, V. R. Burkett, B. Felzer, B. Griffith, et al. *Global Climate Change and Wildlife in North America.* Washington, D.C.: The Wildlife Society, 2004. www.nwf.org/news.

Jablonski, L. M., X. Wang, and P. S. Curtis. "Plant Reproduction under Elevated CO_2 Conditions: A Meta-Analysis of Reports on 79 Crop and Wild Species. *New Phytologist* 156 (2002): 9–26.

Kalkstein, Laurence S. "Direct Impacts in Cities." *Lancet* 342 (December 4, 1993): 1397–1400.

Kiesecker, Joseph M., Andrew R. Blaustein, and Lisa K. Belden. "Complex Causes of Amphibian Population Declines." *Nature* 410 (April 5, 2001): 681–684.

Korner, Christian. "Slow In, Rapid Out—Carbon Flux Studies and Kyoto Targets. *Science* 300 (May 23, 2003): 1242–1243.

LaDeau, Shannon L., and James S. Clark. "Rising CO_2 Levels and the Fecundity of Forest Trees." *Science* 292 (April 6, 2001): 95–98.

Laurance, William F., Alexandre A. Oliveira, Susan G. Laurance, Richard Condit, Henrique E. M. Nascimento, Ana G. Sanchez-Torin, et al. "Pervasive Alteration of Tree Communities in Undisturbed Amazonian Forests." *Nature* 428 (March 11, 2004): 171–175.

Lavers, Chris. *Why Elephants Have Big Ears.* New York: St. Martin's Press, 2000.

Lawton, R. O., U. S. Nair, R. A. Pielke Sr., and R. M. Welch. "Climatic Impact of Tropical Lowland Deforestation on Nearby Montane Cloud Forests." *Science* 294 (October 19, 2001): 584–587.

Lazaroff, Cat. "Climate Change Threatens Global Biodiversity." Environment News Service, February 7, 2001. http://ens-news.com/ens/feb2002/2002L-02-07-06.html.

———. "Loggerhead Turtle Sex Ratio Raises Concerns." Environment News Service, December 18, 2002. http://ens-news.com/ens/dec2002/2002-12-18-06.asp.

Martens, Pim. "How Will Climate Change Affect Human Health?" *American Scientist* 87 (6) (November/December 1999): 534–541.

Martens, Willem J. M., Theo H. Jetten, and Dana A. Focks. "Sensitivity of Malaria, Schistosomiasis, and Dengue to Global Warming." *Climatic Change* 35 (1997): 145–156.

Maslin, Mark, and Stephen J. Burns. "Reconstruction of the Amazon Basin Effective Moisture Availability over the Past 14,000 Years." *Science* 290 (December 22, 2000): 2285–2287.

Mason, John, Jack A. Bailey, and Ardea London. "Doomsday for Butterflies as Britain Warms Up; Dozens of Native Species at Risk of Extinction as Habitats Come under Threat." *London Independent*, September 29, 2002, 12.

Maugh, Thomas H., II. "Global Warming Altering Mosquito." *Los Angeles Times*, November 12, 2001, A-18.

Maxwell, Fordyce. "Climate Warning for Scotland's Wildlife." *The Scotsman*, November 14, 2001, 7.

McEwen, Bill. "The West's Dying Forests." Letter to the Editor. *New York Times*, August 2, 2004, A-16.

McFarling, Usha Lee. "Study Links Warming to Epidemics: The Survey Lists Species Hit by Outbreaks and Suggests That Humans Are Also in Peril." *Los Angeles Times*, June 21, 2002, A-7.

————. "A Tiny 'Early Warning' of Global Warming's Effect: The Population of Pikas, Rabbit-Like Mountain Dwellers, Is Falling, a Study Finds." *Los Angeles Times*, February 26, 2003, A-17.

McMichael, A. J. *Planetary Overload: Global Environmental Change and the Health of the Human Species.* Cambridge: Cambridge University Press, 1993.

McWilliams, Brendan. "Study of Plants Confirms Global Warming." *Irish Times*, November 1, 2001, 26.

Meagher, John. "Look What the Changing Climate Dragged in . . ." *Irish Independent,* July 9, 2004. (Lexis).

Meek, James. "Global Warming Gives Pests Taste for Life in London." *London Guardian*, October 8, 2002, 6.

————. "Tropical Travellers: A Seahorse in the Thames." *London Guardian*, December 12, 2000, 4.

————. "Wildflowers Study Gives Clear Evidence of Global Warming." *London Guardian*, May 31, 2002, 6.

Menzel, A., and P. Fabian. "Growing Season Extended in Europe." *Nature* 397 (1999):659–662.

"More Carbon Dioxide Could Reduce Crop Value." Environment News Service, October 3, 2002. http://ens-news.com/ens/oct2002/2002-10-03-09.asp#anchor2.

Moss, Stephen. "Casualties." *London Guardian*, April 26, 2001, 18.

Munro, Margaret. "Global Warming Affecting Squirrels' Genes, Study Finds: Research in Yukon: 'Phenomenal Change' Seen as Rodents Breeding Earlier." *Canada National Post*, February 12, 2003, A-2.

————. "Puffin Colony Threatened by Warming: A Few Degrees Can Be Devastating. Thousands of Triangle Island Chicks Die When Heat Drives Off Their Favoured Fish." *Montreal Gazette*, July 15, 2003, A-12.

Naik, Gautam, and Geraldo Samor. "Drought Spotlights Extent of Damage in Amazon Basin." *Wall Street Journal*, October 21, 2005, A-12.

Nuttall, Nick. "Strange Visitor Traced to Africa." *London Times*, December 11, 2000. (Lexis).

Ogle, Andy. "Squirrels Get Squirrelier Earlier: Climate Change to Blame. Breeding Season in Yukon Advances 18 Days in Decade." *Edmonton Journal* (Canwest News Service) in *Montreal Gazette*, February 12, 2003, A-12.

O'Malley, Brendan. "Global Warming Puts Rainforest at Risk." *Cairns (Australia) Courier-Mail*, July 24, 2003, 14.

Oren, R., D. S. Ellsworth, K. H. Johnsen, N. Phillips, B. E. Ewers, C. Maier, et al. "Soil Fertility Limits Carbon Sequestration by Forest Ecosystems in a CO_2-Enriched Atmosphere." *Nature* 411 (May 24, 2001): 469–472.

Parmesan, Camille, and Gary Yohe. "A Globally Coherent Fingerprint of Climate Change Impacts across Natural Systems." *Nature* 421 (January 2, 2003): 37–42.

Pegg, J. R. "Plants Prospering from Climate Change." Environment News Service, June 5, 2003. http://ens-news.com/ens/jun2003/2003-06-06-10.asp.

Peng, Shaobing, Jianliang Huang, John E. Sheehy, Rebecca C. Laza, Romeo M. Visperas, Xuhua Zhong, et al. "Rice Yields Decline with Higher Night Temperature from Global Warming." *Proceedings of the National Academy of Sciences* 101 (27) (July 6, 2004): 9971–9975.

Pianin, Eric. "A Baltimore without Orioles? Study Says Global Warming May Rob Maryland, Other States of Their Official Birds." *Washington Post*, March 4, 2002, A-3.

Polakovic, Gary. "Deforestation Far Away Hurts Rain Forests, Study Says: Downing Trees on Costa Rica's Coastal Plains Inhibits Cloud Formation in Distant Peaks. 'It's Incredibly Ominous,' a Scientist Says." *Los Angeles Times*, October 19, 2001, A-1.

Pounds, J. Alan. "Climate and Amphibian Decline." *Nature* 410 (April 5, 2001): 639–640.

Pounds, J. Alan, Martin R. Bustamante, Luis A. Coloma, Jamie A. Consuegra, Michael P. L. Fogden, Pru N. Foster, Enrique La Marca, Karen L. Masters, Andres Merino-Viteri, Robert Puschendorf, Santiago R. Ron, G. Arturo Sanchez-Azofeifa, Christopher J. Still, and Bruce E. Young. "Widespread Amphibian Extinctions from Epidemic Disease Driven by Global Warming." *Nature* 439 (January 12, 2006): 161–167.

———, and Robert Puschendorf. "Clouded Futures." *Nature* 427 (January 8, 2004): 107–108.

Prigg, Mark. "Despite All the Heavy Rain, That Was the Hottest June for 28 Years." *London Evening Standard*, July 1, 2004, A-9.

Radford, Tim. "World Sickens as Heat Rises: Infections in Wildlife Spread as Pests Thrive in Climate Change." *London Guardian*, June 21, 2002, 7.

Reany, Patricia. " 'Millions Will Die' unless Climate Policies Change." Reuters, November 6, 1997. http://benetton.dkrz.de:3688/homepages/georg/kimo/0254.html.

Recar, Paul. "Study: Elements Can Stunt Plant Growth." Associated Press On-line, December 5, 2002. (Lexis).

Reed, Nicholas. "Mild Winter Stirs Wildlife to Early Thoughts of Love." *Vancouver Sun*, February 12, 2003, B-1.

Reynolds, James. "Earth Is Heading for Mass Extinction in Just a Century." *The Scotsman*, June 18, 2003, 6.

Rogers, David J., and Sarah E. Randolph. "The Global Spread of Malaria in a Future, Warmer World." *Science* 289 (September 8, 2000): 1763–1766.

Rohter, Larry. "Deep in the Amazon Forest, Vast Questions about Global Climate Change." *New York Times*, November 4, 2003. www.nytimes.com/2003/11/04/science/earth/04AMAZ.html.

———. "A Record Amazon Drought, and Fear of Wider Ills." *New York Times*, December 11, 2005. www.nytimes.com/2005/12/11/international/americas/11amazon.html.

Root, Terry L., Jeff T. Price, Kimberly L. Hall, Stephen H. Schneider, Cynthia Rosenzweig, and J. Alan Pounds. "Fingerprints of Global Warming on Wild Animals and Plants." *Nature* 421 (January 2, 2003): 57–60.

Ryskin, G. "Methane-Driven Oceanic Eruptions and Mass Extinctions." *Geology* 31 (2003): 737–740.

Schmid, Randolph E. "Warming Climate Reduces Yield for Rice, One of World's Most Important Crops." Associated Press, June 28, 2004. (Lexis).

Schneider, Stephen H. *Global Warming: Are We Entering the Greenhouse Century?* San Francisco: Sierra Club Books, 1989.

———, and Terry L. Root. *Wildlife Responses to Climate Change*. Washington, D.C.: Island Press, 2001.

Seabrook, Charles. "Amphibian Populations Drop." *Atlanta Journal-Constitution*, October 15, 2004, 1-C.

Siegert, F., G. Ruecker, A. Hinrichs, and A. A. Hoffmann. "Increased Damage from Fires in Logged Forests during Droughts Caused by El Niño." *Nature* 414 (November 22, 2001): 437–440.

Sierra Club. "Global Warming: The High Costs of Inaction." 1999. www.sierraclub.org/global-warming/resources/innactio.htm.

Smith, Lewis. "Falling Numbers Silence Cuckoo's Call of Spring." *London Times*, March 6, 2002. (Lexis).

Smith, Stephen. "Comin' Ah-choo: Tepid Temperatures Speeding Allergy Season." *Boston Globe*, April 10, 2002, A-1.

Smucker, Philip. "Global Warming Sends Troops of Baboons on the Run: Rising Temperatures and Humans Encroaching on Grasslands Are Endangering the Ethiopian Primates." *Christian Science Monitor,* June 15, 2001, 7.

Souder, William. "Global Warming and a Toad Species' Decline." *Washington Post,* April 9, 2001. (Lexis).

Stark, Mike. "Assault by Bark Beetles Transforming Forests: Vast Swaths of West Are Red, Gray, and Dying. Drought, Fire Suppression, and Global Warming Are Blamed." *Billings Gazette* in *Los Angeles Times,* October 6, 2002, B-1.

Stein, Rob, and Shankar Vedantam. "Science: Notebook." *Washington Post,* November 12, 2001, A-9.

Stewart, Fiona. "Climate Change in the Back Garden." *The Scotsman,* November 20, 2002, 8.

Stokstad, Erik. "Global Survey Documents Puzzling Decline of Amphibians." *Science* 306 (October 15, 2004): 391.

Stuart, Simon N., Janice S. Cranson, Neil A. Cox, Bruce E. Young, Ana S. L. Rodrigues, Debra L. Fischman, et al. "Status and Trends of Amphibian Declines and Extinctions Worldwide." *Science* 306 (December 3, 2004): 1783–1786.

"Suffocation Suspected for Greatest Mass Extinction." NewScientist.com News Service, September 9, 2003. www.newscientist.com/news/news.jsp?id=ns99994138.

Surendran, Aparna. "Fossil Fuel Cuts Would Reduce Early Deaths, Illness, Study Says." *Los Angeles Times,* August 17, 2001, A-20.

Tangley, Laura. "Greenhouse Effects: High CO_2 Levels May Give Fast-Growing Trees an Edge." *Science* 292 (April 6, 2001): 36–37.

Tanner, Lawrence H., John F. Hubert, Brian P. Coffey, and Dennis P. McInerney. "Stability of Atmospheric CO_2 Levels across the Triassic/Jurassic Boundary." *Nature* 411 (June 7, 2001): 675–677.

Taubes, Gary. "Apocalypse Not." 1997. www.junkscience.com/news/taubes2.html.

Thomas, Chris D., Alison Cameron, Rhys E. Green, Michael Bakkenes, Linda J. Beaumont, Yvonne C. Collingham, et al. "Extinction Risk from Climate Change." *Nature* 427 (January 8, 2004): 145–148.

Toner, Mike. "Warming Rearranges Life in Wild." *Atlanta Journal and Constitution,* January 2, 2003, 1-A.

Unwin, Brian. "Rose in Sightings of Exotic Sea Life Enchant Devon and Cornwall: Fascinating Foreigners Drawn to the Cornish Riviera." *London Independent,* July 15, 2002, 7.

Utton, Tim. "The Rat Rampage." *London Daily Mail,* January 22, 2003. (Lexis).

Verburg, Piet, Robert E. Hecky, and Hedy Kling. "Ecological Consequences of a Century of Warming in Lake Tanganyika." *Science* 301 (July 25, 2003): 505–507.

Wagner, Angie. "Debate over Causes Aside, Warm Climate's Effects Striking in the West." Associated Press, April 27, 2004. (Lexis).

Walther, Gian-Reto. "Plants in a Warmer World." *Perspectives in Plant Ecology, Evolution, and Systematics* 6 (3) (2003): 169–185.

———. "Weakening of Climatic Constraints with Global Warming and Its Consequences for Evergreen Broad-Leaved Species." *Folia Geobotanica* 37 (2002): 129–139.

———, Eric Post, Peter Convey, Annette Menzel, Camille Parmesan, Trevor J. C. Beebee, et al. "Ecological Responses to Recent Climate Change." *Nature* 416 (March 28, 2002): 389–395.

"Warming Climate Pushes Plants to Bloom Earlier, Study Shows." *Wall Street Journal Online*, May 30, 2002. http://phoenix.liunet.edu/~uroy/externalities/earlybloom.html.

Warren, M. S., J. K. Hill, J. A. Thomas, J. Asher, R. Fox, B. Huntley, et al. "Rapid Responses of British Butterflies to Opposing Forces of Climate and Habitat Change." *Nature* 414 (November 1, 2001): 65–69.

Watson, Robert T. Chair, Intergovernmental Panel on Climate Change. Climate Change 2001. Speech at the resumed Sixth Conference of Parties to the United Nations Framework Convention on Climate Change, July 19, 2001. www.ipcc.ch/present/COP65/COP-6-bis.htm.

Webb, Jason. "Mosquito Invasion as Argentina Warms." Reuters, 1998. http://bonanza.lter.uaf.edu/~davev/nrm304/glbxnews.htm.

Weinburg, Bill. "Hurricane Mitch, Indigenous Peoples and Mesoamerica's Climate Disaster." *Native Americas* 16 (3/4) (Fall/Winter 1999): 50–59. http://nativeamericas.aip.cornell.edu/fall99/fall99weinberg.html.

Weiner, Jonathan. *The Next One Hundred Years: Shaping the Fate of Our Living Earth.* New York: Bantam Books, 1990.

Wilmsen, Steven. "Critters Enjoy a Baby Boom: Mild Winter's Downside Is Proliferation of Vermin." *Boston Globe*, March 30, 2002, B-1.

Yoon, Carol Kaesuk. "Something Missing in Fragile Cloud Forest: The Clouds." *New York Times*, November 20, 2001, F-5.

Yozwiak, Steve. "'Island' Sizzle, Growth May Make Valley an Increasingly Hot Spot." *The Arizona (Phoenix) Republic*, September 25, 1998. www.sepp.org/reality/arizrepub.html.

Zavaleta, Erika S., M. Rebecca Shaw, Nona R. Chiariello, Harold A. Mooney, and Christopher B. Field. "Additive Effects of Simulated Climate Changes, Elevated CO_2, and Nitrogen Deposition on Grassland Diversity." *Proceedings of the National Academy of Sciences* 100 (13) (June 24, 2003): 7650–7654.

VI SOLUTIONS

INTRODUCTION

Changing the Energy Paradigm

Having surveyed evidence of worldwide warming, along with the science supporting it and political controversy attending the issue, what, one might ask, are the chances that humankind may dodge this bullet? Chances of forestalling seriously debilitating climate change revolve around humankind's ability to forge political and technological solutions. One will not work without the other. Technological changes range from the very prosaic (such as mileage improvements on existing gasoline-burning automobiles, changes in building codes, and painting building roofs white) to the exotic, including the invention of microorganisms that eat carbon dioxide and the generation of microwaves from the moon. In between are the makings of a shift in the energy paradigm by the end of the century from fossil fuels to renewable, nonpolluting sources such as solar and wind power. By the end of this century, the internal combustion engine may be as much an antique piece as a horse-and-buggy is today.

S. Pacala and R. Socolow, writing in *Science*, have asserted that, using existing technology, "humanity already possesses the fundamental scientific, technical, and industrial know-how to solve the carbon and climate problem for the next half-century" (Pacala and Socolow 2004,

968). By "solve," they mean that the tools are at hand to meet global energy needs without doubling preindustrial levels of carbon dioxide. Their "stabilization strategy" involves intense attention to improved automotive fuel economy, reduced reliance on cars, more efficient building construction, improved power plant efficiency, substitution of natural gas for coal, storage of carbon captured in power plants as well as hydrogen and synthetic fuel plants, more use of nuclear power, development of wind and photovoltaic (solar) energy sources, creation of hydrogen from renewable sources, and more intense use of biofuels such as ethanol. The strategy also advocates more intense management of natural sinks, including reductions in deforestation and aggressive management of agricultural soils through measures such as conservation tillage—drilling seeds into soil without plowing (pp. 969–971).

Martin I. Hoffert and colleagues believe that "mid-century primary-power transmissions could be several times what we now derive from fossil fuels (about 10 to the 13th power watts), even with improvements in energy efficiency" (Hoffert et al. 2002, 981). Hoffert and colleagues' survey of future energy sources includes terrestrial solar and wind energy, solar power satellites, biomass, nuclear fission, nuclear fusion, fission-fusion hybrids, and fossil fuels from which carbon has been sequestered. To this mix they add efficiency improvements, hydrogen production, storage and transport, superconducting global electric grids, and geo-engineering (p. 981). This effort will challenge the creative resources of scientists as much as the Apollo project that put men on the moon. In addition, the scientists said, the use of fossil fuels must decline, a matter of political decisionmaking that exceeds the reach of science.

Research must begin immediately to build an energy infrastructure that is "climate-neutral." Without such action, the atmosphere's concentration of greenhouse gases will double from preindustrial levels by the end of the twenty-first century, scientists say. "A broad range of intensive research and development is urgently needed to produce technological options that can allow both climate stabilization and economic development" (Revkin [November 1] 2002, A-6). This group of eighteen researchers called for intensive new efforts to improve existing technologies and develop others, such as fusion reactors or space-based solar power plants. The researchers provided only the broadest estimate of how much this research effort will cost. "The cost probably will total tens of billions of dollars in both government and private funds" (p. A-6).

Most existing energy technologies "have severe deficiencies," the scientists said. Solar panels, new nuclear power options, windmills, filters for fossil fuel emissions, and other options are either inadequate or require vastly more research and development than is currently planned in the United States or elsewhere (Revkin [November 1] 2002, A-6). One author of the analysis, Dr. Haroon S. Kheshgi, a chemical engineer with ExxonMobil, said that "climate change is a serious risk" requiring a shift away from fossil fuels. "You need a quantum jump in technology," he said. "What we're talking about here is a 50- to 100-year time scale" (p. A-6).

Hoffert, a New York University physics professor, said he was convinced that technological problems will be solved on a scientific and engineering level. At the same time, he was worried whether the public and its elected officials will realize the urgency of the task (Revkin [November 1] 2002, A-6). Several of the study's authors said that, at present, they see few signs that major industrial nations were ready to engage in an ambitious quest for climate-neutral energy. Richard L. Schmalensee, a climate policy expert and the dean of the Massachusetts Institute of Technology Sloan School of Management, said the issue of climate change remained too complex and contentious to generate the requisite focus. "There is no substitute for political will," he said (p. A-6). Some environmental advocates criticized this study's emphasis on still-distant technologies, asserting that this focus could distract from the need to do what is possible at present to reduce greenhouse gas emissions. "Techno-fixes are pipe dreams in many cases," said Kert Davies, research director for Greenpeace, which has been conducting a broad campaign against ExxonMobil. "The real solution," he said, "is cutting the use of fossil fuels by any means necessary" (p. A-6). Given a sufficient sense of urgency, however, Hoffert believed that technology will solve the problem once a large proportion of the world's people and its leaders realize the seriousness of global warming. "We started World War II with biplanes," he commented, "and seven years later we had jets" (Revkin [November 3] 2002, A-8).

The United States could reduce its greenhouse gas emissions without sacrificing economically, according to a study by the World Wildlife Fund (WWF). Using a combination of lower fuel and electricity bills due to more efficient appliances, buildings, and automobiles, the United States could more than pay the costs of developing new technology and putting it into place, resulting in savings of $135 billion annually in energy costs by 2020, according to the study. By 2010, U.S. households would save an average of $113 each, and the nation could cut carbon dioxide emissions to a level about 2.5 percent above that of 1990, this study said. Jennifer Morgan, director of the WWF's climate-change campaign, asserted: "What this basically does is disprove the president's [George W. Bush's] claim and shows rather that the treaty could help the United States meet its energy goals. The U.S. would be much better off in such an agreement. This study shows that if you're serious about climate change, you can also wean yourself off foreign oil and make your economy more efficient" (Heilprin 2001).

Myron Ebell, global-warming director for the Competitive Enterprise Institute, dismissed the WWF study assertions. He said a nation relying on the burning of fossil fuels for 80 percent of its energy needs cannot simply turn on a dime. "They are playing with fantasy assumptions and numbers," he said. "There is a way we could comply with the Kyoto treaty cheaply, and that is to lower our standard of living substantially. Energy is part of our standard of living, it's part of our wealth" (Heilprin 2001).

Although scientists can warn us that change in our energy paradigm is necessary to stem damaging climate changes, they cannot change it for us. By the early twenty-first century, as international debate stalled the Kyoto Protocol, a number of other avenues of change opened. The United Kingdom's leaders, for example, began to increase diplomatic pressure on a recalcitrant U.S. federal government. Increasing numbers of public protests focused attention on climate-change issues. Several U.S. states began to legislate in areas where the federal government lagged. A great deal of talk, and some actions, promoted climate-change solutions in many large corporations in oil, automobiles, and electric utilities, all key to how energy is produced and used.

Many possible solutions are being developed and debated, including fertilization of the oceans with iron (to promote growth of carbon dioxide–consuming algae), improvements in farming technology, "sequestration" (burial on land or in the oceans) of human-generated carbon dioxide, as well as strategies to limit emissions of methane.

20 ACTING GLOBALLY AND LOCALLY

LEADERSHIP FROM THE UNITED KINGDOM

During mid-May 2002, Britain's government launched a strong attack on George W. Bush's rejection of the Kyoto Protocol, warning that such actions ultimately threatened to make large parts of the Earth uninhabitable. Great Britain's environment minister, Michael Meacher, wrote in the *London Guardian* that the world is running out of time. "We do not have much time and we do not have any serious option. If we do not act quickly to minimize runaway feedback effects [from global warming] we run the risk of making this planet, our home, uninhabitable" (Watt 2002, 11).

Meacher spoke after the U.S. chief climate negotiator, Harlan Watson, said in London that an independent U.S. initiative to cut emissions of greenhouse gases would not be assessed until 2012. "We are not going to be part of the Kyoto protocol for the foreseeable future," he announced (Watt 2002, 11). "Climate change may be not steady but abrupt," rejoined Meacher. "The pressures we inflict on the climate may trigger wholly unexpected developments from feedback effects." Latest scientific evidence suggests the impact of climate change on Britain could be "faster and sharper" than expected, said Meacher. Almost 2 million homes in England and Wales are at risk from floods, and Britain will experience a 65 percent increase in river flooding if flood-control defenses do not take climate change into account (p. 11). Between 1990 and

2002, Great Britain reduced carbon dioxide emissions 15 percent while increasing industrial output 30 percent—proof that the Bush administration was wrong when it asserted that abiding by the terms of the Kyoto Protocol would ruin the economy.

Even the leader of Britain's opposition Conservative Party, Michael Howard, has criticized President George W. Bush for failing to tackle climate change. Immediately after the Conservatives' disparaging remarks, British Prime Minister Tony Blair pressured the United States and Russia to face up to the "catastrophic consequences" of climate change, as he issued his starkest warning yet about the "alarming and unsustainable" consequences of global warming. He said that within the lifetimes of his children—and possibly in his own lifetime—the impact on the world could be so far-reaching and "irreversible in its destructive power" that it will alter human existence radically (Jones and Clover 2004, 2). By October 2004, even Britain's Queen Elizabeth was criticizing Bush's inertia on the subject.

PUBLIC PROTESTS ESCALATE OVER ENERGY POLICY INERTIA

Public protests of energy policy inertia have become more common in both the United States and Europe. On May 15, 2003, for example, a coalition of environmental and public-interest groups dumped a ton of coal on the lawn of the U.S. Capitol to protest subsidies to that industry. In England during the third Saturday in May 2002, the world's largest oil conglomerate, ExxonMobil, found its 400 service stations besieged by thousands of protesters who objected to the company's lack of initiative on global warming and urged motorists to take their business elsewhere. Farmers posted roadside "Stop Esso" billboards, and the company was condemned at several local music festivals. Using another tactic, about fifty bicyclists on June 12, 2004, rode through downtown Seattle in the nude, wearing only helmets and body paint, to protest society's continued dependence on fossil fuels. During the fall of 2004, after Florida was devastated by four major hurricanes, a coalition of scientists and environmentalists erected billboards showing a satellite image of a menacing hurricane off the state's coast and proclaiming: "Global warming equals worse hurricanes. George Bush just doesn't get it" (Royse 2004).

On another front, during 2002, the world's largest oil company tried to silence its biggest critic by taking Greenpeace to court over the use of ExxonMobil logos in a "Stop Esso" boycott campaign. The Texas-based energy group accused Greenpeace of damaging its reputation by doctoring logo letters to resemble the "SS" moniker of the Nazi secret police. ExxonMobil demanded that Greenpeace pay it 80,000 euros for damage of its reputation, as well as a further 80,000 euros a day should it continue to use the offending material. Greenpeace

Greenpeace launched a "Stop Global Warming" balloon over the Mae Moh coal-fired power plant in Lampang, Thailand. Courtesy of Greenpeace/Unanongrak.

commented that the legal action was a signal that the world's largest oil company was being hurt by the campaign launched to highlight Exxon's position on global warming (Macalister 2002, 21).

"French law protects your trademark and logo and our employees and customers would not understand if we did not take action to prevent its misuse," said an ExxonMobil spokeswoman in its French office, where the suit had been filed. Exxon's court documents said that the company aimed to prevent publication of symbols on the Stop Esso internet site. The middle two letters of Esso have been replaced by dollar signs, not the Nazi "SS" logo, said Greenpeace. "Instead of using bully-boy antics to gag free speech, we suggest Esso instead halt its campaign to subvert international action on climate change," said Stephen Tindale, Greenpeace United Kingdom director. "We simply replaced two letters in Esso's logo with the internationally recognized symbol for the U.S. dollar. We find it ironic that the richest corporation in the world can't recognize the dollar sign and confuses it with a Nazi symbol," he added (Macalister 2002, 21).

ExxonMobil, on July 8, 2002, won the first round of the increasingly bad-tempered legal battle designed to halt Greenpeace's Stop Esso campaign. Exxon was awarded an injunction forcing Greenpeace to cease use of its Esso logo with dollar signs scrawled through it, pending a full hearing. The environmental group was allowed four working days to remove the offending logo from its website or face a fine of 5,000 euros.

In the United States, a number of Greenpeace activists on May 27, 2003, blocked the entrance of ExxonMobil's headquarters at Irving, Texas, to protest company inaction vis-à-vis global warming. Thirty-six of the protesters were arrested by Irving police and charged with criminal trespassing, a misdemeanor. Many were dressed in tiger costumes, mimicking the old Exxon mascot Tony the Tiger. The protest was staged one day before the company's annual shareholders' meeting in Dallas. At that meeting, environmental activists faced off with pro-Exxon demonstrators who sang "Give Oil a Chance," a parody of John Lennon's "Give Peace a Chance."

Physical confrontations between English Greenpeace activists and the oil industry have become frequent. On February 16, 2005, the day the Kyoto Protocol came into force, thirty-five Greenpeace activists wearing business suits invaded the International Petroleum Exchange in London, the world's second-largest energy market, shutting down trading for an hour before several were arrested. The market exchange, which sets the price for more than half the world's crude oil, was invaded about 2 p.m. by activists blaring foghorns, alarms, and whistles in an attempt to disrupt "open-cry" floor trading. Some of them attached distress alarms to helium balloons that they lofted above the trading floor. Security guards kicked and punched activists, wounding twelve of them, as three members of Greenpeace scaled the exchange building to hang a banner proclaiming, "Climate Change Kills—Stop Pushing Oil" (Vidal and Macalister 2005, 4).

U.S. STATES TAKE ACTION

Several U.S. states and cities, no longer waiting for the George W. Bush administration to seize the initiative against global warming, have begun taking steps to reduce greenhouse gas emissions. By 2005, 166 U.S. cities had agreed to meet the goals of the Kyoto Protocol, which the Bush administration had rejected (*Harper's* Index 2005, 11).

By late 2005, officials in nine northeastern states had decided to ask their legislatures to control carbon dioxide emissions from more than 600 power plants within their borders, in defiance of the federal government's refusal to take similar action. California, Washington, and Oregon also were in the early stages of exploring a regional agreement similar to the northeastern plan. The nine states in the northeastern agreement are Connecticut, Delaware, Maine, Massachusetts, New Hampshire, New Jersey, New York, Rhode Island, and Vermont.

Portland, Oregon has made reduction of greenhouse gas emissions a municipal priority; by 2005, emissions there had been reduced to roughly 1990 levels as the city's economy has boomed. Portland has encouraged walking and bicycle commuting, and told local companies that if they gave employees free parking they should also subsidize bus passes. One step (replacing bulbs in traffic lights with light-emitting diodes) cut electrical use by 80 percent and saved the city almost $500,000 a year in electrical costs (Kristof 2005). Officials in Portland said that they have been able to cut emissions there in accordance with Kyoto while enjoying a healthy economy, with "less money spent on energy, more convenient transportation, a greener city, and expertise in energy efficiency that is helping local business win contracts worldwide" (Daynes and Sussman 2005, 442).

Portland is one of 158 U.S. cities whose mayors in June 2005 signed the U.S. Mayors Climate Protection Agreement, which commits the cities to trim greenhouse gas emissions in their jurisdictions to levels close to those required by the Kyoto Protocol.

State and local governments are legislating "broader use of energy-saving devices, more energy-efficient building standards, cleaner-burning power plants and more investment in such renewable energy sources as wind and solar power" (Polakovic 2001, A-1). By 2002 more than half the states in the United States had adopted voluntary or mandatory programs for reducing carbon dioxide emissions, according to the Pew Center on Global Climate Change. Fifteen states, including President George W. Bush's home state of Texas, have enacted legislation requiring utilities to increase their use of renewable energy sources such as wind power or biomass in generating electricity (Pianin 2002, A-3).

California, for example, enacted regulations to reduce car and truck emissions by 2006 that, if they survive a challenge expected from the auto industry, could be a model for New York, New Jersey, and other northeastern states. Texas has legally required that 3–4 percent of its electricity come from renewable energy sources, notably wind power, by 2010. Massachusetts became the first state legally to limit power plant emissions of carbon dioxide by six major coal- and oil-burning facilities in that state as part of its rules controlling pollution. At about the same time, New Hampshire enacted emission controls for that state's three aging coal-fired electric generating plants. Nebraska became the first state to enact legislation to require rotation of crop planting, increasing the number of plants and trees that absorb carbon dioxide.

During June 2005, California Governor Arnold Schwarzenegger broke ranks with the Bush administration, advocating wide-ranging curbs on carbon dioxide emissions in what, considered as a separate nation, would be the sixth-largest economy in the world and its tenth-largest source of greenhouse gases. Schwarzenegger advocated curbs on economic grounds to prevent reduction of water supplies, rising sea levels, and increases in agricultural pests. He called for lowering emissions to 2000 levels by 2010, and to 1990 levels by 2020. By 2050, Schwarzenegger proposed an

California Governor Arnold Schwarzenegger delivers a speech during a bill signing ceremony at the United Nations World Environment Day June 1, 2005, in San Francisco, California. He signed Executive Order S-3-05, which established greenhouse gas emission reduction targets for California. © Justin Sullivan/Getty Images.

80 percent reduction compared to 1990. "I say the debate is over," the governor said. "We know the science. We see the threat. And we know the time for action is now" (Stokstad 2005, 1530).

New Jersey during 1999 set a voluntary goal of reducing total greenhouse gases to 3.5 percent below 1990 levels by 2005. A charge was added to consumers' utility bills that raised $358 million for energy efficiency and renewable energy programs (Pianin 2002, A-3). With Governor John Baldacci's signature, Maine during 2002 became the first state to set legal goals for overall reduction of carbon dioxide emissions. The legislation required Maine to develop a climate-change action plan to reduce carbon dioxide to 1990 levels by 2010, 10 percent

645

below 1990 levels by 2020, and by as much as 75–80 percent over the long term. "My legislative colleagues recognize the impact climate change will have on Maine's economy, environment, and quality of life," said Representative Ted Koffman of Bar Harbor, lead sponsor of the bill. "The challenges posed by global warming will be especially felt by future generations," Koffman said. "The breakthrough legislation signed today will bring together private and public interests to collaboratively develop a cost effective plan to reduce global warming while saving energy in the process" ("Maine" 2003).

During August 2001, six New England states and five eastern Canadian provinces signed a pact to reduce greenhouse gas emissions. Signatories pledged to cut emissions to 1990 levels by 2010 and by 10 percent below that level by 2020, goals similar to those of the Kyoto Protocol. Those cuts were to be followed in future years by even deeper reductions. This is, according to a report in the *Los Angeles Times*, "the most ambitious goal set by state governments, . . . supported by three Republican governors, two Democrats and one Independent from Massachusetts, Connecticut, Rhode Island, New Hampshire, Vermont and Maine" (Polakovic 2001, A-1).

According to a survey by the Environment News Service, in December 2001, New Hampshire issued a list of policy options for achieving the goals established by the New England Governors-Eastern Canadian Premiers Climate Action Plan. The state still needs to develop a plan with specific reductions and a concrete timeline. Massachusetts during 2003 drafted a climate-action plan designed to meet or exceed the regional goals established by the New England governors and eastern Canadian premiers. In July 2002, Rhode Island published a climate action plan designed to meet the greenhouse gas emission targets of the climate plan. In August 2002, the governor of Vermont issued an executive order establishing a goal of reducing that state's greenhouse gas emissions by more than 25 percent over the next decade, consistent with the regional goals ("Maine" 2003).

Many of these agreements seek to increase the efficiency of electricity generation in the United States. The U.S. electricity sector emits nearly twice as much carbon dioxide per megawatt-hour of energy as that produced in Europe. In 2001, U.S. plants produced an average of 720 kilograms (1,600 pounds) of carbon dioxide per megawatt-hour of electricity production, compared to 353 kilograms for European companies. Part of the disparity results from a greater dependence on nuclear plants to produce electricity in Europe—33 percent of its production compared to 20 percent in the United States. Fossil fuel–burning plants in Europe also generate 10–25 percent more energy per unit of carbon dioxide than similar U.S. installations ("U.S. Electricity" 2002).

"This is a way [for states] to distinguish themselves from the Bush administration," California Resources Secretary Mary Nichols said. "A lot of practical, moderate people are recognizing [that] climate change is a reality, not a theory, and they need to take it into account and help move the direction of the world by doing something about it" (Polakovic 2001, A-1). California has been an early leader in reducing dependence on fossil fuels that release greenhouse gases. By 2001, 12 percent of California's electrical power came from renewable sources, more than any other state (p. A-1).

New York Governor George Pataki in June 2001 announced a series of measures to improve energy efficiency and trim greenhouse gases. Specifically, Pataki ordered state buildings to acquire 20 percent of their electricity from renewable sources, such as solar or wind power, by 2010. Meanwhile, in Oregon, more than 20,000 people signed up for the "Blue Sky Program" by paying an extra $3 per month on their utility bills to ensure that Pacific Power and Light purchases electricity from sources that don't contribute to global warming.

During early June 2003, Connecticut, Maine, and Massachusetts sued the federal government's Environmental Protection Agency in an attempt to force classification of carbon dioxide emissions as air pollution for regulatory purposes vis-à-vis global warming. The suit asserted that it was the first by U.S. states to compel action on climate change. If the suit succeeds, the EPA will be required to classify carbon dioxide as a "criteria pollutant" under the federal Clean Air Act. According to the states' lawyers, this classification would trigger establishment of standards for allowable levels in the atmosphere, as the federal government now does for ozone, lead, sulfur dioxide, and other gases. Accusing the federal government of neglecting the threat of global warming, during July 2004, eight states and New York City sued some of the largest U.S. power companies, asking for legal action that would force them to reduce their carbon dioxide emissions.

U.S. STATES ACT ON AUTOMOBILE EFFICIENCY

The California Air Resources Board, defying the auto industry, voted unanimously during late September 2004 to approve the world's most stringent rules reducing automobile emissions. Under the regulations, the automobile industry must cut exhaust from cars and light trucks by 25 percent and from larger trucks and sport-utility vehicles by 18 percent. The industry will have until 2009 to begin introducing cleaner technology and will have until 2016 to meet the new exhaust standards.

California's plan for sharp cuts in automotive emissions of greenhouse gases could eventually lead most states on the East and West coasts of the United States to require similar emissions cuts. In turn, these requirements may provoke the automakers to adopt the same standards for cleaner, more fuel-efficient vehicles across their model lines. The only way to cut global warming emissions from cars is to use less fossil fuel. Because of this limitation, proposed cuts in legally allowable emissions would, as a side effect, force automakers to increase fuel economy by roughly 35 to 45 percent. California's plan, as proposed in 2004, requires automakers to cut greenhouse gas emissions in their new vehicles by 29.2 percent over a decade, phasing in gradually from the 2009 to the 2015 model years.

During 2004, the governments of New Jersey, Rhode Island, and Connecticut said that they intended to follow California's automobile rules instead of the federal government's. New York, Massachusetts, Vermont, and Maine already had adopted the California rules. "Let's work to reduce greenhouse gases by adopting the carbon-dioxide emission standards for motor vehicles which were recently proposed by the State of California," New York Governor George E. Pataki said in his state-of-the-state address in 2003. These seven states and California account for almost 26 percent of the U.S. auto market, according to R. L.

Polk, a company that tracks automobile registrations (Hakim [June 11] 2004, C-4). Automakers from Detroit to Tokyo believe that these states, along with Canada, could form a potent bloc requiring decreases in the emissions of greenhouse gases from automobiles. "It would be a logistical and engineering challenge, and a costly problem," said Dave Barthmuss, a spokesman for General Motors. "It's more cost-effective for us to have one set of emissions everywhere." "If they only want to make one car," said Roland Hwang, a senior policy analyst at the Natural Resources Defense Council, "clearly it should be a clean car, and that's the California car (p. C-4).

MANY U.S. COMPANIES SLOW TO TAKE ACTION ON CLIMATE CHANGE

Unlike many of their foreign rivals, American industry giants such as ChevronTexaco, ExxonMobil, General Electric, Southern Company, and Xcel Energy after the year 2000 continued to pursue business strategies that discounted the global-warming threat, according to a report by the Investor Responsibility Research Center (IRRC). "Such strategies leave them and their shareholders especially vulnerable to the increased financial risks and missed market opportunities posed by climate change," said Doug Cogan, author of the study and deputy director of social issues for IRRC. "Companies cannot expect to mitigate climate-change risks and seize new market opportunities until they build a foundation of well functioning environmental management systems and properly focused governance practices for a carbon-constrained world," Cogan asserted ("Many U.S." 2003).

The report, "Corporate Governance and Climate Change: Making the Connection," was commissioned by CERES, a coalition of investor, environmental, and public-interest groups, and compiled by IRRC, an independent organization that advises institutional investors managing more than $5 trillion in assets. The report profiled twenty companies. These included the top five carbon dioxide producers in electric power, auto, and petroleum industries, as well as five other industry leaders. The report used a Climate Change Governance Checklist to analyze the

Toyota Motor Corporation President Fujio Cho introduced the company's newly developed hybrid SUVs, the Harrier Hybrid (left) and the Kluger Hybrid (right), during a press conference on March 22, 2005, in Tokyo, Japan. © Koichi Kamoshida/Getty Images.

companies' actions in the areas of board oversight, management accountability, executive compensation, emissions reporting, and material-risk disclosure ("Many U.S." 2003).

The twenty companies, all of which are widely held by institutional investors, included Alcoa, American Electric Power, British Petroleum, ChevronTexaco, Cinergy, ConocoPhillips, DaimlerChrysler, DuPont, ExxonMobil, Ford Motor Company, General Electric, General Motors, Honda, IBM, International Paper, Royal Dutch/Shell, Southern, Toyota, TXU, and Xcel Energy. According to this study, the U.S. electric power industry scored lowest on the checklist, despite its position as the largest single source of greenhouse gas emissions and its vulnerability to changing clean-air regulations. The auto industry also failed to measure and disclose emissions of its products. At the same time, Japanese competitors have been taking the lead in introducing hybrid gas-electric vehicles that substantially reduce tail-pipe emissions ("Many U.S." 2003).

The IRRC report found the largest differences in corporate-governance responses to climate change within the oil industry. British Petroleum and Royal Dutch/Shell have pursued all fourteen items listed on the Climate Change Governance Checklist, whereas American-based rivals ChevronTexaco, ConocoPhillips, and ExxonMobil have pursued only four or five actions. Unlike their foreign counterparts, the U.S.-based oil companies continued to devote nearly all of their development efforts to fossil fuels, largely ignoring renewable energy technologies. DaimlerChrysler, General Electric, and TXU are other companies with low scores, having taken only four or five actions. Alcoa and DuPont stood out among the U.S. companies profiled, having pursued twelve of the fourteen actions ("Many U.S." 2003).

CORPORATE RESPONSIBILITY ON GLOBAL WARMING: TALKING AND WALKING

Talk is cheap—and no cheaper than when it comes from motor-company and oil-industry executives doing lip service on global warming. Are executives who talk the talk also walking the walk? In the talking-the-talk department, a blue ribbon goes to Atlantic Richfield Oil's chief executive, Mike Bowlin, who gave a speech to an oil-industry audience in January 1999 in which he said that his industry is now entering "the last days of the age of oil." Oil and gas companies face a crucial choice, he said. "Embrace the future and recognize the growing demand for a wide array of fuels; or ignore reality and slowly but surely be left behind" (Leggett 2001, 329).

Witness also Chief Executive Officer William Clay Ford of the Ford Motor Company, great-grandson of Henry Ford. William Clay Ford at one point assembled a well-financed corporate division that promoted alternatives to the internal-combustion engine, including hydrogen fuel cells, electric commuter cars, and even bicycles. "The global temperature is rising and the evidence suggest that the shift is being affected by human activity, including emissions related to fossil fuels used for transportation," Ford told Paul McKay of the *Ottawa Citizen*. "We believe it is time to take appropriate action" (McKay 2001, A-1). Other automakers also have paid at least lip service to the issue. Daimler-Chrysler's chairman, Juergen Schrempp, has said that he supports the goals of the Kyoto Protocol. The burning of gasoline by automobiles is the second-largest source of carbon dioxide emissions in the United States, after power plants.

During the first week of October 2000, Ford, speaking before a Greenpeace business conference in London, anticipated the demise of the internal-combustion engine. During that speech, Ford described an all-out race by automakers to design the first mass-produced hydrogen fuel-cell vehicle, anticipating that major Japanese and American car makers would be using fuel-cell vehicles to supersede gasoline engines by the year 2003 and 2004. At that time, according to one newspaper account, Ford called global warming "the most challenging issue facing the world." Furthermore, said Ford, "I believe fuel cells will finally end the 100-year reign of the internal-combustion engine." Ford continued, "The climate appears to be changing, the changes appear to be outside natural variation, and the likely consequences will be serious. From a business-planning point of view, that issue is settled. Anyone who disagrees is, in my view, still in denial" (McCarthy 2000, 10).

Ford Motor earned some notice among environmentalists for its annual "corporate citizenship reports," which included frank discussion of subjects such as the impact of sport-utility vehicles (SUVs) on global warming. Ford's first corporate-responsibility report, issued in 2000, took issue with the company's rising production of gas-thirsty, carbon-belching SUVs. In its second "corporate citizenship report" (2001), Ford said that it had formed a committee of executives to examine ways to reduce the company's contribution to global warming. "On the issue of climate change, there's no doubt that sufficient evidence exists to move from argument to action," Jacques Nasser, Ford's chief executive, said in a preamble to the report (Bradsher 2001, C-3). The 2002 report, Ford's third, was released at a time of financial distress for the company. In that report, Ford said, "difficult business conditions make it harder to achieve the goals we set for ourselves in many areas, including corporate citizenship. . . . But that doesn't mean we will abandon our goals or change our direction" (Hakim 2002, C-4).

In the talk-the-talk department, Ford committed itself to a 25 percent improvement in gas mileage for its sport-utility vehicles in five years, a goal that, by 2004, had not even been seriously breathed upon in that part of reality where the rubber meets the road. Lip service is about all environmentalists got from Ford; gas mileage on its 2003 model SUVs was worse than for the 2002s. Still, Ford executives were very good at dreaming fuel efficiency. On August 4, 2004, for example, the company's highest executives, including William Clay Ford, gathered at corporate headquarters. They came away, according to a *New York Times* report two months later, glowing to each other about a projected 80 percent improvement in fuel efficiency by the year 2030. The *Times* report added that "the company had not planned to publicize the strategy because it is a long-term objective subject to change and because the company has recently been under attack from environmental groups for falling short of previously stated environmental objectives" (Hakim [October 5] 2004, 11).

The *Montreal Gazette* complained editorially that the automobile industry seemed intent on selling the car-buying public ever-larger SUVs, no matter what the IPCC and the industry's own spin-doctors said about their effects on climate.

> It's easy to mistake the latest automotive news for a joke. Consumers next year will be able to buy a passenger vehicle that is a meter longer than today's largest-sport utility vehicle and twice—repeat, twice—as heavy. It sounds like a joke, but it's true—and all too symptomatic of North America's breezy insouciance toward gas guzzling and climate change. Appropriately, the vehicle's maker will be Freightliner, the DaimlerChrysler subsidiary best known for turning out 18-wheelers. . . . The vehicle's ad slogan resonates: "You don't need roads, when you can make your own." It perfectly captures the ethos that nature should not get in the way. ("Global Warning" 2001, B-2)

A minority of shareholders at some companies have become restive with regard to climate change. Shareholders, for example, have filed global-warming resolutions at General Motors and Ford annual meetings that attempt to increase pressure on the automakers to reduce their greenhouse gas emissions. A coalition of shareholders—mostly members of various Catholic orders—asked that the automakers report carbon dioxide emissions from their plants and vehicles. "We believe that both General Motors and Ford face material and reputational risk in their current failure to address and reduce carbon dioxide emissions," said Sister Patricia Daly, executive director of the Tri-State Coalition for Responsible Investment (Eggert 2002).

A resolution calling for a report on how ExxonMobil will respond to the growing pressure to develop renewable energy won support from 21 percent of shareholders at the ExxonMobil annual general meeting May 28, 2003. The 21 percent shareholder support for the renewables resolution represented $42.34 billion worth of ExxonMobil stock ("One in Five" 2003). The same day, almost a quarter of Southern Company shareholders (one of the largest U.S. utilities) voted to require the company to evaluate potential financial risks associated with its use of fossil fuels. Earlier, 27 percent of shareholders at American Electric Power (AEP) Company had voted to support a similar resolution. Southern, AEP, Excel Energy Company, Cinergy Company, and TXU Corporation are the top five emission sources of carbon dioxide in the United States. A similar resolution at ChevronTexaco won 32 percent of shareholders' votes in 2003, up from 9.6 percent in 2001 (Seelye 2003, C-1).

ExxonMobil pledged $10 million a year for ten years to Stanford University for climate research, most of it to help establish a research center that will develop energy technologies that do not add to greenhouse gas emissions. ExxonMobil, at the same time, has been maintaining sizable donations to groups that question global warming's scientific basis, including right-wing lobbying organizations such as the Competitive Enterprise Institute, Frontiers of Freedom, George C. Marshall Institute, American Council for Capital Formation Center for Policy Research, and American Legislative Exchange Council.

SOME BUSINESSES ACCEPT AND PROFIT FROM CARBON CURBS

In some cases, private businesses in the United States have taken actions ahead of the federal government. "We accept that the science on global warming is overwhelming," said John W. Rowe, chairman and chief executive officer of Exelon Corporation. "There should be mandatory carbon constraints" (Carey and Shapiro 2004). Exelon, the United States' largest operator of commercial nuclear power plants, probably would benefit from such constraints.

Shell Company's chairman, Sir Philip Watts, has called for global warming skeptics to accept action to limit greenhouse gas emissions "before it is too late" (Macalister 2003, 19). Watts stated that "we can't wait to answer all questions [on global warming] beyond reasonable doubt," adding, "there is compelling evidence that climate change is a threat" (p. 19). Watts made his remarks at the opening of a new Shell Center for Sustainability at Rice University in Houston. Executives at Shell have "seen and heard enough" to believe that the burning of fossil fuels poses a problem, Watts said. "We stand with those who are prepared to take action to solve that problem...now...before it is too late...and we believe that businesses, like Shell, can help to bridge differences that divide the U.S. and Europe on this issue" (p. 19).

Shell Oil has developed "an impressive roster of alternative energy projects" through a $500 million investment in a subsidiary, Shell Renewables. Shell constructed a four-megawatt wind farm off the coast of northeastern England that was supplying 3,000 homes by 2001. BP (British Petroleum) Solar, by 2000, became the largest solar-power company in the world. In 2000, it sold forty megawatts of new capacity. ExxonMobil, on the other hand, spent $500 million on renewable energy and then "decided that the practical [obstacles to] turning renewables into profitable energy sources were too great to warrant further funding" (Schrope 2001, 516).

American Electric Power (AEP) Company, which burns more coal than any other U.S. electric utility, once resisted the idea of combating climate change. During the late 1990s, however, then Chief Executive Officer E. Linn Draper Jr. pushed for a strategy shift, preparing for limits instead of denying that global warming existed. "We felt it was inevitable that we were going to live in a carbon-constrained world," said Dale E. Heydlauff, AEP's senior vice president for environmental affairs (Carey and Shapiro 2004). AEP has invested in renewable energy projects in Chile, retrofitted school buildings in Bulgaria for greater efficiency, and explored ways to burn coal more cleanly.

Many other companies have also been acting. Late in 2000, the Du-Pont Company, once the world's largest producer of ozone-consuming chloroflourocarbons (CFCs), announced that an ambitious corporate program had eliminated half of the company's greenhouse gas emissions compared to 1990, nearly all of it without losing sales or profits. By 2004, DuPont had cut its greenhouse gas emissions by 65 percent since 1990, saving hundreds of millions of dollars (Carey and Shapiro 2004). Florida Power & Light Company by 2004 had forty-two wind-power facilities and had promoted energy efficiency, reducing emissions and eliminating the need to build ten mid-sized power plants, according to Randall R. LaBauve, vice president for environmental services (Carey and Shapiro 2004). British Petroleum (whose advertising styles "BP" as "Beyond Petroleum") reduced its greenhouse gas emissions 10 percent and saved $650 million during three years by performing a detailed energy inventory of its operations. Private U.S. companies in 2004 formed a Climate Group to share information about climate-related business planning, from new technologies to plugging leaks.

IS THE KYOTO PROTOCOL A BAND-AID OR A DEAD LETTER?

Global greenhouse gas emissions are rising, and evidence of a warming planet is accumulating much more quickly than world diplomacy has been able to address the situation. The snail-paced nature of consultative diplomacy combines with the reality that we feel the results of fossil fuel effluvia perhaps forty years after the fact (through a complex set of natural feedbacks) to create a trap in which human responses to global warming take place several decades after nature requires them.

Given these circumstances, the Kyoto Protocol may be a climatic dead letter, even though its approval by Russia in September 2004 produced worldwide implementation on paper. Russia joined 124 other countries in ratifying the protocol and, with its 17.5 percent share of worldwide carbon dioxide emissions, raised the total signatories to

David Merrill, executive director of the National Global Warming Coalition, read a paper urging U.S. President George W. Bush to sign the Kyoto Protocol during a protest on the Ellipse behind the White House on February 14, 2005, in Washington, D.C. © Mark Wilson/Getty Images.

cover slightly more than 60 percent of worldwide emissions, above the 55 percent required to bring Kyoto into force.

Seven years after its negotiation in 1997, however, the only sizable countries that have come close to meeting Kyoto Protocol target emission reductions have been Great Britain and Germany. Most other signatories have not met their goals, and most third world countries (India and China among them) are not bound by its provisions. The Kyoto Protocol has become more of a political rallying cry than a serious challenge to global warming. Even if the protocol is fully implemented, a projected temperature rise of 2 degrees C by 2050 would be shaved only by 0.07 degree C, according to calculations by atmospheric scientist Thomas M. L. Wigley. In other words, the Kyoto goals are only a small fraction of the reduction in emissions required if worldwide temperature levels are to be stabilized during the twenty-first century and afterward (Wolf 2000, 27).

As governments around the world dickered over climate diplomacy (and the United States, which produces about one-fifth of the world's greenhouse gases, ignored the Kyoto Protocol), global emissions of carbon dioxide from fossil fuel combustion increased by 13 percent above 1990 levels by the year 2000, mainly due to large pollution increases in developing nations and substantial growth in the United States and other Western industrialized nations, according to statistics compiled by the International Energy Agency. The increase would have been higher, except for the collapse of former state socialist economies in Russia and eastern Europe during the period (Holly 2002). Carbon dioxide emissions for the period rose by 17.8 percent in the United States, from 4.8 billion tons in 1990 to 5.7 billion tons in 2000, while western European emissions rose by 3.9 percent. As a result of the dissolution of the Soviet Union and the resulting economic collapse in former Soviet nations and eastern Europe, carbon dioxide emissions in these nations fell from 3.7 billion tons in 1990 to 2.6 billion tons, a drop of 30.6 percent (Holly 2002).

Emissions rose in all major economic sectors, including energy, transport, industry, and agriculture. One exception was waste management, where emissions declined slightly. The figures did not include emissions and removals from land-use change and forestry. Greenhouse gas emissions in the highly industrialized countries as a whole rose by 8 percent from 1990 to 2000. Emissions increased in most other highly industrialized countries, including 5 percent in New Zealand, 11 percent in Japan, 18 percent in Australia, and 20 percent in Canada ("Rich Countries" 2003).

Emissions of carbon dioxide and other greenhouse gases from Europe, Japan, the United States, and other industrialized countries could grow by 17 percent from 2000 to 2010, despite measures in place to curb them, according to a U.N. report. "These findings clearly demonstrate that stronger and more creative policies will be needed for accelerating the spread of climate-friendly technologies and persuading businesses, local governments and citizens to cut their greenhouse gas emissions," said Joke Waller Hunter, executive secretary of the U.N. Climate Change Convention ("Rich Countries" 2003).

EUROPE'S GREENHOUSE GAS EMISSIONS

Changes in European countries' greenhouse gas emissions varied dramatically between 1990 and 1999. Spain's emissions rose 23.2 percent, and Ireland's increased 22.1 percent, as the Netherlands increased 6.1 percent, Italy 4.4 percent, and Denmark 4.0 percent. Greenhouse gas emissions of the European Union overall declined by 8.0 percent, led by Germany (down 18.7 percent) and aided by the collapse of inefficient industries in the former East Germany. Great Britain's greenhouse gas emissions declined 12.8 percent during the decade (Winestock 2001, A-9).

Emissions of carbon dioxide fell by 5.3 percent between 1990 and 2001 in Britain; emissions of carbon dioxide for 2002 were estimated to have been 8–9 percent lower than in 1990. Total greenhouse gas emissions in 2002 were estimated to have been between 14 and 15 percent below the 1990 level. Britain's reductions resulted mainly from reduced energy consumption per unit of economic output, warmer weather, and a decrease in use of coal relative to oil and gas. The British government said that the United Kingdom aims to move beyond its Kyoto target toward its goal of reducing emissions of carbon dioxide by 20 percent below 1990 levels by 2010, and 60 percent by 2050 (A. Brown 2003).

To stop further damage to the global climate, a 60 percent reduction in emissions by 2050 is required, United Kingdom Prime Minister Tony Blair said February 24, 2003, in a major speech on sustainable development. Speaking at an event organized by the U.N. Sustainable Development Commission, Blair advocated an international consensus to tackle key issues of sustainable development, including climate change. Blair said that the Kyoto Protocol would at best deliver a 2 percent reduction in greenhouse gas emissions worldwide (P. Brown 2003, 13). With his speech, Blair's administration released an Energy White Paper advocating a switch from greenhouse fuels toward wind, wave, and solar power. The alternative, said the report, was temperatures in London as warm as those of Madrid and a risk of tornadoes and other storms with hurricane-force winds. Rising seas also could submerge some coastal cities, as temperatures rise by as much as 6 degrees C (Ingham and Keevins 2003, 17).

John Prescott, British deputy prime minister, during 2002 announced a £69 million package of measures to combat global warming, including payments to motorists who switch to more efficient transportation. Owners of cars that can be converted to use liquid propane gas in addition to gasoline will be able to claim up to 75 percent of the cost of conversion up to £2,500. Buyers of such cars will be eligible for grants of £1,000 or more. Other programs subsidize the conversion of gasoline-driven cars to electric power.

REFORESTATION AND GLOBAL WARMING: PROSPECTS AND PROBLEMS

Carbon uptake from reforestation has been proposed to reduce net carbon dioxide emissions to the atmosphere, even as some models indicate that forests sometimes contribute to global warming, on balance. For example, "The albedo of a forested landscape is generally lower than that of cultivated land, especially when snow is lying. . . . In many boreal forest areas, the positive forcing induced by decreases in albedo can offset the negative forcing that is expected from carbon sequestration" (Betts 2000, 187). According to this analysis, high-latitude reforestation efforts actually might worsen the greenhouse effect.

Other research indicates that older, wild forests are far better than plantations of young trees at removing carbon dioxide from the atmosphere. One such analysis, published in the journal *Science*, was completed by Dr. Ernst-Detlef Schulze, the director of the Max Planck Institute for Biogeochemistry in Jena, Germany, and two other scientists at the institute. The study provided an important new argument for protecting old-growth forests. The scientists said that their study provides a reminder that the main goal should be to reduce carbon dioxide emissions at the source—at smokestacks and tail pipes.

The German study, together with other similar research, has produced a picture of mature forests that differs sharply from long-held notions in forestry, Schulze said. He said that aging forests were long perceived to be in a state of decay that releases as much carbon dioxide as it captures. Soils in undisturbed tropical rainforests, Siberian woods, and some German national parks contain enormous amounts of carbon derived from fallen leaves, twigs, and buried roots that can bind to soil particles and remain stored for 1,000 years or more. When such forests are cut, the trees' roots decay, and soil is disrupted, releasing the carbon dioxide (Revkin [September 22] 2000, A-23; Schulze, Wirth, and Heimann 2000). "In contrast to the sink management proposed in the Kyoto Protocol, which favors young forest stands, we argue that preservation of natural old-growth forests may have a larger effect on the carbon cycle than promotion of re-growth" (Schulze, Wirth, and Heimann 2000, 2058). Instead of reducing the level of carbon dioxide in the atmosphere, the Kyoto Protocol's emphasis on new growth at the expense of established forests "will lead to massive carbon losses to the atmosphere mainly by replacing a large pool with a minute pool of regrowth and by reducing the flux into a permanent pool of soil organic matter. Both effects may override the anticipated aim, namely to increase the terrestrial sink capacity by afforestation and reforestation" (p. 2059).

"In old forests, huge amounts of carbon taken from the air are locked away not only in the tree trunks and branches, but also deep in the soil, where the carbon can stay for many centuries," said Kevin R. Gurney, a research scientist at Colorado State University. When such a forest is cut, he said, almost all of that stored carbon is eventually returned to the air in the form of carbon dioxide. "It took a huge amount of time to get that carbon sequestered in those soils," he stated. "So if you release it, even if you plant again, it'll take equally long to get it back" (Revkin [September 22] 2000, A-23).

OCEAN FERTILIZATION WITH IRON:
PROMISE AND PERIL

Should the oceans be seeded with large amounts of iron ore that will stimulate the growth of carbon dioxide–consuming phytoplankton? The idea has attracted some support among corporations and foundations looking for ways to minimize the effects of carbon dioxide without changing the world's basic energy-generation mix. The idea is simple at first glance: iron stimulates the growth of phytoplanktonic algae, which is believed to be responsible for about half of the world's biologic absorption of carbon dioxide. Ulf Riebesell, a marine biologist at the Alfred Wegener Institute for Polar and Marine Research in Bremerhaven, Germany, believes that ambitious iron seeding of the oceans could remove 3 to 5 billion tons of carbon dioxide per year, or about 10–20 percent of human-generated emissions, and be at least an order of ten times cheaper than planting forests to do the same thing (Schiermeier 2003, 110). Patents have been issued for ocean fertilization, and demonstration projects undertaken. One such project has been described and evaluated in *Nature* (Boyd et al. 2000, 695–702; Watson et al. 2000, 730–733).

Nearly half of the Earth's photosynthesis is performed by phytoplankton in the world's seas and oceans. This fact has led to proposals to "geo-engineer" a carbon dioxide sink via the same "biological pump" that is believed to have driven at least some of the Earth's past climate cycles (Chisholm 2000, 685). The chemistry of the oceans varies widely with regard to the amount of iron necessary to prime this pump and substantially increase carbon dioxide sequestration. Sallie W. Chisholm, a marine biologist at the Massachusetts Institute of Technology, wrote in *Nature*, it "is possible to stimulate the productivity of hundreds of square kilometers of ocean with a few barrels of fertilizer" (p. 686).

In an experiment conducted between Tasmania and Antarctica, researchers confirmed that vast stretches of the world's southern oceans are primed to explode with photosynthesis but lack only iron. The researchers, who described their work in *Nature*, said it is too soon to start large-scale iron seeding, because the new experiment raised as many questions as it answered. At best, they said, iron seeding would absorb only a small amount of the carbon dioxide in the atmosphere. They also said that their experimental bloom of plankton was not tracked long enough to determine whether the carbon harvested from the air sank into the deep sea or was again released into the environment as carbon dioxide gas. Atsushi Tsuda and colleagues have studied iron fertilization and have found that, under some circumstances, it can dramatically increase phytoplankton mass (Tsuda et al. 2003, 958–961). "There are still fundamental scientific questions that need to be addressed before anyone can responsibly promote iron fertilization as a climate control tactic," said Kenneth H. Coale, an oceanographer who has helped design studies of iron's effects in the tropical Pacific (Revkin [October 12] 2000, A-18).

This seemingly simple proposition has some potential problems, however. First, there exists no way to measure the amount of carbon taken up by phytoplankton. Additionally, the algae produce dimethyl-sulphide, which plays a role in cloud formation; phytoplankton increase the amount of sunlight absorbed by ocean water, as well as heat energy. They also produce compounds such as methyl halides, which play a role in stratospheric ozone depletion. The iron also could promote the growth of toxic algae that might kill other marine life and change the chemistry of ocean water by removing oxygen. "The oceans are a tightly linked system, one part of which cannot be changed without resonating through the whole system," said Chisholm. "There is no free lunch" (Schiermeier 2003, 110).

So much iron may be required to produce the desired effect that fertilization of this type will never be commercially possible. "The experiments enabled us to make an initial determination about the amount of iron that would be required and the size of the area to be fertilized," said Ken O. Buesseler of the Woods Hole Oceanographic Institution, who coauthored a study of the idea.

> Based on the studies to date, the amount of iron needed and [the] area of ocean that would be impacted is too large to support the commercial application of iron to the ocean as a solution to our greenhouse gas problem. It may not be an inexpensive or practical option if what we have seen to date is true in further experiments on larger scales over longer time spans. The oceans are already naturally taking up human-produced carbon dioxide, so the changes to the system are already underway. We need to first ask will it work and then what are the environmental consequences? ("Iron Link" 2003)

One study asserts that "to assess whether iron fertilization has potential as an effective sequestration strategy, we need to measure the ratio of iron added to the amount of carbon sequestered in the form of particulate organic carbon to the deep ocean in field studies" (Buesseler and Boyd 2003, 67). To date, wrote Buesseler and Boyd, experiments of this type have "produced notable increases in biomass and associated decreases in dissolved inorganic carbon and macronutrients. However, evidence of sinking carbon particle carrying P.O.C. [particulate organic carbon] to the deep ocean was limited" (p. 67). Given the limits of present technology, this study estimates that an area larger than the southern ocean (waters southerly of 50 degrees south latitude) would have to be fertilized to remove 30 percent of the carbon that human activity presently injects into the atmosphere. Thus, according to this study, "ocean iron fertilization may not be a cheap and attractive option if impacts on carbon export and sequestration are as low as observed to date" (p. 68).

Despite prospective problems, iron fertilization is considered possibly viable in some quarters. Scientists who fed tons of iron into the southern ocean reported evidence during 2004 that stimulating the growth of phytoplankton in this way may strengthen the oceans' use as a carbon sink. In a report published April 16, 2004, in *Science*, ocean biologists and chemists from more than twenty research centers said they triggered two huge blooms of phytoplankton that turned the ocean green for weeks and consumed hundreds, perhaps thousands, of tons of carbon dioxide. "These findings would be encouraging to those considering iron fertilization as a global geoengineering strategy," said Ken Coale, a chief scientist at the Moss Landing Marine Laboratories. "But the scientists involved in this experiment realize that this looked only skin deep at the functioning of ocean ecosystems and much more needs to be understood before we recommend such a strategy on a global scale." "From my work, I don't think this could solve a significant fraction of our greenhouse-gas problem while causing unknown ecological consequences," said Buesseler (Hoffman 2004).

BUILDING CODE CHANGES

Considerable mitigation of global warming may be accomplished by wise use of new technology to improve the energy efficiency of dwellings, factories, and offices. Energy consumption of heating and air conditioning systems could be reduced by as much as 90 percent in new buildings, for example, with modern insulation, triple-glazed windows with tight seals, and passive solar design (Speth 2004, 65).

On January 1, 2003, Australia changed its national building code with the explicit purpose of reducing energy consumption. Amendment 12 of the Building Code Australia includes a range of measures appropriate for different climate zones of Australia that address wall, ceiling, floor, and glazing thermal performance to avoid or reduce the use of energy for artificial heating and cooling. This reduction in energy use is achieved by utilizing passive solar heating where appropriate; using natural ventilation and internal air movement to avoid or reduce the use of artificial cooling; sealing houses in some climates to reduce energy loss through leakage; insulating to reduce heat loss from water piping of central heating systems; and insulating and sealing to reduce energy loss through the walls of ductwork associated with heating and air conditioning systems.

CARBON DIOXIDE'S USE AS A REFRIGERANT

Several hundred researchers from around the world met at Purdue University July 12–15, 2004, to discuss innovative air conditioning and refrigeration technologies, including designs aimed at reducing global warming. At least one of the ideas, the use of carbon dioxide as a refrigerant, is nearly a century old. Carbon dioxide was the first refrigerant used during the early twentieth century but was later replaced with man-made chemicals. Carbon dioxide may be on the verge of a comeback because of technological advances that include the manufacture of extremely thin aluminum tubing called "microchannels" ("Conferences Tackle" 2004).

Hydrofluorocarbons, today's most widely used refrigerants, cause about 1,400 times more global warming than the same quantity of carbon dioxide. The HFCs replaced chlorofluorocarbons (CFCs), also potent greenhouse gases that degrade stratospheric ozone. Tiny quantities of carbon dioxide released from air conditioners would be insignificant compared to the huge amounts produced from burning fossil fuels for energy and transportation, said Eckhard Groll, an associate professor of mechanical engineering at Purdue ("Conferences Tackle" 2004).

Carbon dioxide offers few advantages for large air conditioners, which do not have space restrictions and can use wide-diameter tubes capable of carrying enough of the conventional refrigerants to provide proper cooling capacity. Carbon dioxide, however, may be a promising alternative for systems that must be small and lightweight, such as automotive or portable air conditioners. Various factors, including the high operating pressure required for carbon dioxide systems, enable the refrigerant to flow through small-diameter tubing, which allows engineers to design more compact air conditioners ("Conferences Tackle" 2004).

In the meantime, Greenpeace asked refrigerator manufacturers to use hydrocarbons such as propane instead of HFCs and developed its own "Greenfreeze" technology. Unilever became interested in the idea as part of its commitment to reduce the impact of its activities on climate change. The Greenfreeze technology was adequate for domestic refrigerators, and Unilever carried out lengthy trials to create a system that could be used in larger-scale freezers as well. Unilever agreed to trials at the 2000 Olympics in Sydney and committed to a global change-over to HFC-free refrigerants.

IMPROVEMENTS IN FARMING TECHNOLOGY

Rattan Lal, a professor of soil science at Ohio State University, has asserted that the atmosphere's load of carbon dioxide could be reduced markedly by several relatively simple changes in farming technology. Contributions of farming to carbon dioxide in the atmosphere have been increasing with rising populations. Carbon dioxide is added to the atmosphere via plowing, so Lal believes that reducing the depth of furrows would significantly reduce the amount of carbon dioxide introduced into the atmosphere by agriculture.

During mid-2003, the U.S. Department of Agriculture announced plans to give incentives to farmers for management practices that keep carbon in the soil. For the first time, the U.S. Department of Agriculture began to factor reduction of greenhouse gas emissions into soil-conservation programs by giving priority to farmers who reduce emissions of carbon dioxide, methane, and nitrous oxides. Such programs represented $3.9 billion in federal spending during the 2003–2004 fiscal year. Farmers were encouraged to use no- or low-tillage methods, as well as crop rotation, buffer strips, and other practices that reduce greenhouse gas emissions as well as soil erosion. Such practices were expected to retain 12 million tons of greenhouse gases by 2012 (Clayton 2003, D-1).

Farming with an eye to carbon sequestration would utilize soil restoration and woodland regeneration, no-till farming, cover crops, nutrient management, manuring and sludge application, improved grazing, water conservation, efficient irrigation, agroforestry practices, and growth of energy crops on spare lands. Intensive use of such practices, according to one estimate, could "offset fossil-fuel emissions by 0.4 to 1.2 gigatons of carbon per year, or 5 to 15 per cent of the global fossil-fuel emissions" (Lai 2004, 1623).

Tim O'Riordan of the Zuckerman Institute for Connective Environmental Research said, "We have to put sustainable development at the heart of businesses such as fish farming and agriculture. We need agricultural stewardship schemes that have incentives for farmers to produce according to sustainable principles, which in turn will deliver healthy soil, water and wild life. This, in turn, should offer jobs in recreation and education for eco-care. We also need the involvement of the local community to ensure that all acts of stewardship have neighborhood understanding and support" (Urquhart and Gilchrist 2002, 9).

DEEP-SEA INJECTION OF CARBON DIOXIDE: EFFECTS ON LIFE

Disposal of carbon dioxide in the deep oceans has been proposed as one method of mitigating global warming. However, many proposals to inject human-generated carbon dioxide into the oceans ignore the possible effects of such sequestration on life at these levels. Brad A. Seibel and Patrick J. Walsh examined these effects, finding that increased deep-water carbon dioxide levels result in decreases of seawater pH (increasing acidity), which can be harmful to sea creatures, "as has been demonstrated for the effects of acid rain on freshwater fish" (Seibel and Walsh 2001, 319). They find that "a drop in arterial pH by just 0.2 would reduce bound oxygen in the deep-sea crustacean *Glyphocrangon vicaria* by 25 per cent" (p. 320). The same drop in arterial pH would reduce bound oxygen in the mid-water shrimp *Gnathophausia ingens* by 50 percent.

According to Seibel and Walsh, "deep-sea fish hemoglobins are even more sensitive to pH" (2001, 320). Small increases in carbon dioxide levels and resulting decreases of pH levels "may trigger metabolic suppression in a variety of organisms," as "low pH has been shown to inhibit protein synthesis in trout living in lakes rendered acidic through anthropogenic effects" (p. 320). Seibel and Walsh cited research by R. L. Haedrich that "any change that takes place too quickly to allow for a compensating adaptive change within the genetic potential of finely adapted deep-water organisms is likely to be harmful." Seibel and Walsh conclude that "available data indicate that deep-sea organisms are highly sensitive to even modest pH changes.... Small perturbations in CO_2 or pH may thus have important consequences for the ecology of the deep sea" (p. 320).

"Through various feedback mechanisms, the ocean circulation could change and affect the retention time of carbon dioxide injected into the deep ocean, thereby indirectly altering oceanic carbon storage and atmospheric carbon dioxide concentration," said Atul Jain, a professor of atmospheric sciences at the University of Illinois at Urbana-Champaign. "Where you inject the carbon dioxide turns out to be a very important issue" ("Global Warming Could Hamper" 2002). To investigate the possible effects of feedbacks between global climate change, the ocean carbon cycle, and oceanic carbon sequestration, Jain and graduate student Long Cao developed an atmosphere-ocean, climate-carbon cycle model of intermediate complexity. The researchers then used the model to study the effectiveness of oceanic carbon sequestration by the direct injection of carbon dioxide at different locations and ocean depths (2002).

Jain and Cao found that climate change has an important impact on the oceans' ability to store carbon dioxide. The effect was most pronounced in the Atlantic Ocean. "When we ran the model without the climate feedback mechanisms, the Pacific Ocean held more carbon dioxide for a longer period of time," Cao said. "But when we added the feedback mechanisms, the retention time in the Atlantic Ocean proved far superior. Based on our initial results, injecting carbon dioxide into the Atlantic Ocean would be more effective than injecting it at the same depth in either the Pacific Ocean or the Indian Ocean" ("Global Warming Could Hamper" 2002).

Future climate change could affect both the uptake of carbon dioxide in the ocean basins and the ocean circulation patterns themselves. Jain further explained:

As sea-surface temperatures increase, the density of the water decreases and thus slows down the ocean Thermohaline Circulation, so the ocean's ability to absorb carbon dioxide also decreases. This leaves more carbon dioxide in the atmosphere, exacerbating the problem. At the same time, the reduced ocean circulation will decrease the ocean mixing, which decreases the ventilation to the atmosphere of carbon injected into the deep ocean. Our model results show that this effect is more dominating in the Atlantic Ocean. . . . Sequestering carbon in the deep ocean is, at best, a technique to buy time. Carbon dioxide dumped in the oceans won't stay there forever. Eventually it will percolate to the surface and into the atmosphere. ("Global Warming Could Hamper" 2002)

By 2002, tourism promoters, commercial fishermen, environmentalists, and sports groups had united in the Hawaiian Islands to oppose experiments in carbon dioxide sequestration off the island of Kauai. The experiments had drawn support from ChevronTexaco, General Motors, Ford, and ExxonMobil. Opponents fear that carbon dioxide sequestration will increase the acidity of the local ocean water and imperil animal and plant life. The U.S. Department of Energy had allowed the companies to begin work without an environmental impact statement because they had asserted human beings would not be affected. Congressional representative Patsy Mink of Hawaii said that "Hawaii's ocean environment is too precious to put at risk for an experiment of this kind" (Dunne 2002, 2).

At about the same time, an effort was underway to deposit 5.4 tons of pure carbon dioxide deep under the North Sea near Norway. This experiment was set to begin during the summer of 2002, until environmentalists campaigned successfully to stop it. According to an Environment News Service report, "The Norwegian oil firm Statoil already was injecting roughly 1 million tons of CO_2 per year into the rock strata of an offshore oilfield in the North Sea, but no one has yet tried sequestration in the oceans." Led by the Norwegian Institute for Water Research, a coalition including United States, Japanese, Canadian, and Australian organizations was planning to inject five tons of liquid CO_2 at 800 meters depth off the coast of Norway ("Liquid CO_2" 2002). Norway's Pollution Control Agency granted the project a discharge permit in early July 2002, subject to approval by the Environment Ministry.

Environmental groups argued that the project would mean "dumping" carbon dioxide in the ocean in violation of the 1972 London Dumping Convention as well as the 1992 Ospar Convention on Protection of the North Sea Environment. The Ospar Commission discussed this issue in late June 2002. "The sea is not a dumping ground. It's illegal to dump nuclear or toxic waste at sea, and it's illegal to dump CO_2—the fossil fuel industry's waste," said Truls Gulowsen, Greenpeace Norway climate campaigner ("Liquid CO_2" 2002). A last-minute veto from Norway's environment minister, Boerge Brende, on August

26, 2002, stopped the project. "The possible future use of the ocean as a storage place for CO_2 is controversial" and "could violate current international rules concerning sea waters," said Environment Minister Boerge Brende ("Norway Says No" 2002).

LIMITING METHANE EMISSIONS

Unlike carbon dioxide emissions, which have continued to rise, the proportion of methane in the atmosphere stagnated during the 1990s, following a peak about 1980. Methane levels have probably stopped rising because of concerted efforts by governments around the world, according to a report from The Goddard Institute for Space Studies (Hansen and Sato 2001, 14778; "Limiting Methane" 2002). By 2003, methane emissions were poised to decline further, in large part because of more efficient oil production in Russia. During 2003, scientists with Australia's Commonwealth Scientific and Industrial Research Organization (CSIRO) announced that methane levels in the Earth's atmosphere had been stable since 1999, after a 15 percent rise during the preceding twenty years and a 150 percent spike since preindustrial times ("Australian Scientists" 2003).

A landfill. Courtesy of Getty Images/PhotoDisc.

The rate of methane growth has slowed during the past decade, and it may be possible to halt its growth entirely and eventually reduce atmospheric levels, James Hansen and colleagues at the Goddard Institute for Space Studies suggested. On July 29, 2004, the Bush administration announced a new agreement with seven other nations to reduce and capture methane emissions at landfills, coal mines, and oil and gas systems, and to utilize the recovered gas for energy. Another warming agent deserving special attention, according to the authors, is soot, a product of incomplete combustion. Diesel-powered trucks and buses are primary sources of airborne soot in the United States. Even larger amounts of soot are delivered into the atmosphere from developing countries (Hansen and Sato 2001, 14778; "Limiting Methane" 2002).

While methane growth may be slowing, Hansen and colleagues emphasized that carbon dioxide emissions are the single largest factor that has contributed to climate change since 1850. They warned that carbon dioxide emissions must be slowed soon and eventually curtailed to stabilize atmospheric conditions and stop global warming. If fossil fuel use continues at today's rates for the next fifty years, and if growth of methane and air pollution is halted, the warming in fifty years will be about 1.3 degrees F (0.7 C), the GISS study forecast. That amount of warming is significant, according to Hansen's team, but it is less than half the warming in the "business as usual" scenarios of the Intergovernmental Panel on Climate Change (IPCC) (Hansen and Sato 2001, 14778; "Limiting Methane" 2002). "The growth rate of climate forcing by measured greenhouse gases peaked near 1980 at almost 5 Watts/per square meter per century," according to Hansen and Makiko Sato, writing in the *Proceedings of the National Academy of Sciences*. They continued:

This growth rate has since declined to about 3 W/square meter per century, largely because of cooperative international actions. We argue that trends can be reduced to the level needed for the moderate "alternative" climate scenario (about 2 W/Square meter per century for the next 50 years) by means of concerted actions

that have other benefits, but the forcing reductions are not au-
tomatic "co-benefits" of actions that slow carbon-dioxide emissions. Current trends of climate forcings by aerosols remain very
uncertain. Nevertheless, practical constraints on changes in emission levels suggest that global-warming at a rate of plus 0.15 degrees C. plus or minus 0.05 degrees C. will occur over the next
several decades. (p. 14778)

Although Hansen and Sato's forecast temperature rise is more
moderate than that of the IPCC, they noted that, even at this pace,
within a century carbon dioxide levels on the Earth will "approach that
of the middle Pliocene, 2.75 million years ago, when the world was
about 2 degrees C. warmer than today and sea level was at least 25
meters higher" (p. 14783).

Methane has been easier to reduce than carbon dioxide. For example, cattle respiration may account for nearly 20 percent of the
methane gas released into the atmosphere. Researchers at the University of Nebraska are developing a food additive for cattle that may
reduce the amount of methane they generate. "The reason we're fo-
cusing on methane is because it's a short-lived, highly potent greenhouse gas that needs to be reduced," said University of Nebraska at
Lincoln biochemistry professor Stephen Ragsdale (Thiessen 2003).
Methane is not only released from a cow via burping and flatulence but
also through ordinary respiration. The methane is produced in the
cow's rumen, the first of a cow's four stomachs, enters the bloodstream,
and exits through the lungs. Ken Olson, a range livestock nutritionist at
Utah State University, conducted a six-year study that found better
range management practices, such as providing higher-quality forage,
can make a small difference in the amount of methane released by
cattle.

Ragsdale and fellow researchers James Takacs and Jess Miner con-
ceived the idea of reducing methane by blocking the enzymes in the
cow's rumen that are required to produce methane. An additive to
block enzymes has been patented by them. They have tested more than
200 compounds in an attempt to find the right formula that blocks

methane but doesn't harm the beneficial microbes in the cow's rumen. Such a product might be tested on smaller animals before cattle—sheep or even termites, Ragsdale said. The Nebraska researchers said they don't believe methane does anything beneficial for cattle. In fact, the researchers believe up to 16 percent of the feed given to cattle is wasted because it is used to produce methane. If methane production was eliminated, producers could feed cattle less because more of the animal's energy would be going toward production of proteins, amino acids, and fat (Thiessen 2003).

MUSICAL GROUPS AND GREENHOUSE GAS EMISSIONS

In its March 15, 2003, edition, the music industry trade journal *Billboard* published a lengthy front-page article detailing how musical groups have joined with Future Forests to plant trees to make up for the ecological costs of their activities. The *Billboard* article said that "numerous artists and music companies are taking a leading role in an environmental program that aims to combat global warming. Foo Fighters, Coldplay, Gorillaz, Kylie Minogue, Shaggy, Mis-teeq, Dido, Neneh Cherry, and Sting—to name a few acts—have linked with Future Forests, a London-based, for-profit company, to ensure that their activities do not exacerbate the ecological problems facing the planet" (Masson 2003, 1). "There are serious problems storing up for us now," Future Forests founder and chairman Dan Morrell was quoted as saying. "But basically, by planting trees, we can make everything we do carbon-neutral, and that's at least a start in fighting those problems" (p. 1).

Future Forests calculated the levels of carbon dioxide created by such activities as manufacturing and distributing a compact disc or staging a concert. According to its calculations, a specific number of trees can be planted to offset those carbon dioxide emissions, thereby making that activity carbon-neutral. Morrell said that the strategy of using music to reach other businesses and the public has been reaping rewards. "Pink Floyd has three forests around the world, and fans visit those forests," he said. Future Forests' charter pledges that it plants only trees that are indigenous to their locations that will be permitted to grow for at least ninety-nine years. All sites are required to allow public access (Masson 2003, 1).

The Rolling Stones performed a special free concert to raise awareness about global warming on February 6, 2003, at the Los Angeles Staples Center. The Natural Resources Defense Council joined with the Rolling Stones to stage the event, hoping, it said, to turn up the heat on the Bush administration's inaction regarding global warming. The partners also emphasized opportunities to start addressing the problem. A private anonymous donor absorbed the costs of the concert; tickets were awarded in a lottery-style sweepstakes.

BIBLICAL STEWARDSHIP OF THE EARTH

The Bible's content is diverse enough to be quoted in almost any context. The same Good Book that commands us to multiply and subdue the Earth also may be quoted to commend stewardship of the natural world. The U.S. Conference of Catholic Bishops has done so in its new "plea for dialogue, prudence, and the common good," its consensus statement on "global climate change." The statement continues: "How are we to fulfill God's call to be stewards of creation in an age when we may have the capacity to alter that creation significantly, and perhaps irrevocably? We believe our response to global climate change should be a sign of our respect for God's creation" (Catholic Bishops 2001).

The bishops' statement continued:

Global climate is by its very nature a part of the planetary commons. The earth's atmosphere encompasses all people, creatures, and habitats. . . . Stewardship [is] defined in this case as the ability to exercise moral responsibility to care for the environment. . . . Our Catholic tradition speaks of a "social mortgage" on property and, in this context, calls us to be good stewards of the Earth. . . . Stew-Stewardship requires a careful protection of the environment and calls us to use our intelligence "to discover the earth's productive potential and the many different ways in which human needs can be satisfied." (Catholic Bishops 2001, quoting John Paul II)

According to this statement, Catholic response to the challenge of climate change "must be rooted in the virtue of prudence." While some uncertainty remains, it said, most experts agree that "something significant is happening to the atmosphere. Human behavior and activity are, according to the most recent findings of the international scientific bodies charged with assessing climate change, contributing to a warming of the earth's climate." This statement included an ecological twist on beliefs in American exceptionalism: "Because of the blessings God has bestowed on our nation and the power it possesses, the United States bears a special responsibility in its stewardship of God's creation to shape responses that serve the entire human family" (Catholic Bishops 2001).

The statement concluded that responsibility weighs more heavily upon those with the power to act, because the threats are often greatest for those who lack similar power, namely vulnerable poor peoples as well as future generations. According to reports of the IPCC, significant delays in addressing climate change may compound the problem and make future remedies more difficult, painful, and costly. On the other hand, said the bishops, the impact of prudent actions today can potentially improve the situation over time, avoiding more painful but necessary actions in the future (Catholic Bishops 2001).

The bishops believe that passing along the problem of global climate change to future generations as a result of our delay, indecision, or self-interest would be easy. However, the statement said, "We simply cannot leave this problem for the children of tomorrow. As stewards of their heritage, we have an obligation to respect their dignity and to pass on their natural inheritance, so that their lives are protected and, if possible, made better than our own" (Catholic Bishops 2001).

"Grateful for the gift of creation," says this statement, "we invite Catholics and men and women of good will in every walk of life to consider with us the moral issues raised by the environmental crisis. . . . These are matters of powerful urgency and major consequence. They constitute an exceptional call to conversion. As individuals, as institutions, as a people, we need a change of heart to preserve and protect the planet for our children and for generations yet unborn" (Catholic Bishops 2001, quoting *Renewing the Earth*).

REDUCE AIR TRAVEL?

Air travel distances soared a hundredfold during the last half of the twentieth century and are expected to rise substantially during the first half of the twenty-first. The number of kilometers flown by airline passengers doubled worldwide from 125 billion in 1990 to 260 billion in 2000. Atmospheric emissions from aircraft, because they include nitrogen oxides and other greenhouse gases, have three times the global-warming potential of carbon dioxide emitted alone. The chemicals emitted into the atmosphere by combusting aviation fuel "cause the formation of polar stratospheric clouds, affect markedly the aerosol composition of the atmosphere, and intensify the greenhouse effect" (Kondratyev, Krapivin, and Varotsos 2003). The IPCC estimates that commercial jets are responsible for 3.5 percent of man-made global warming. With air travel likely to double by 2015, aviation's contribution to global warming will probably rise significantly (Webster 2001).

The number of passengers passing through British airports increased fivefold between the mid-1970s and 2000 and is expected to more than double again by 2020 (Lean 2001, 23). Carbon dioxide emissions from air traffic originating in the United Kingdom rose 85 percent between 1990 and 2000, according to its Department of Trade and Industry (Houlder [December 16] 2002, 2). According to a report published by the Institute of Public Policy Research, "Flying by jet plane is the least environmentally sustainable way to travel and transport goods" (Lean 2001, 23). The air travel industry has reduced the amount of fuel it burns to transport each of its passengers by half since the mid-1970s, but the growth of air travel has more than balanced these savings in fuel efficiency. The same relationship applies to nitrogen oxides and hydrocarbons.

Tim O'Riordan, of the Zuckerman Institute for Connective Environmental Research at the University of East Anglia, said that "everyone who uses a car or flies regularly should consider what positives they can put back into the environmental battle by way of compensation. Maybe, for example, you already cycle to work, or take public transport, as a chosen alternative to driving. The next time you fly somewhere, either by choice or for business, the carbon dioxide and nitrous oxide production from the plane you use will cancel out a year of contributions from cycling" (Urquhart and Gilchrist 2002, 9).

During late November 2002, two official British studies called for an end to cheap flights and a ban on new airport runways. The Royal Commission on Environmental Pollution urged Britain's government to halt airport growth, raise fares, and place financial pressure on short-haul and no-frills carriers. At the same time, Tony Blair's body of environmental advisers, the Sustainable Development Commission, said that proposals for new runways at Stansted, Heathrow, Luton, Rugby, or a new airport at Cliffe in Kent required a "fundamental rethink." The commission, a body of academics, businessmen, and people from public life, recommended that the price of a one-way ticket should rise by £40 to have any chance of mitigating climate change. Paul Ekins, an economist and a member of the commission, said: "We believe a stable climate is a good thing and worth modifying human behavior for" (Clover and Millward 2002, 1, 4).

Brian Hoskins, a research professor at the Royal Society and former head of meteorology at Reading University, commented that, if growth of air travel continued unrestricted, airline vapor trails would cover 10 percent of the sky over Britain by 2050. Vapor trails form cirrus clouds that contribute to global warming, not to mention the emissions from aircraft engines. The commission called for the rapid growth of high-speed trains to replace short-haul services and for the development of "transport hubs, such as Schipol in the Netherlands, rather than airports (Clover and Millward 2002, 1, 4).

The Boeing Company conceded in 2001 that its new high-speed aircraft will use more fuel than existing airliners but asserted that there was "plenty of fossil fuel still around" (Webster 2001). The Sonic

Cruiser will burn 15–20 percent more fuel than other planes in order to accelerate to just below the speed of sound (760 miles per hour).

Taxes on airline fuel are banned by international treaty (the 1944 Chicago Convention on taxing fuel), giving the industry a unique encouragement to pollute. Emissions from international flights have been excluded from the treaties negotiated to combat global warming and ozone depletion. Airports in Britain are exempt from pollution control, from getting planning permission for many developments, and—in most cases—from statutory noise regulation. And, to top it off, air travel receives billions of dollars every year in subsidies from tax-payers of industrialized countries (Lean 2001, 23).

CARBON TAXES

Late in 2002, Great Britain's Royal Society, the country's leading scientific body, called on the government to abolish its climate-change levy, a tax on businesses' use of energy. The Royal Society advocated replacement of the levy with a carbon tax or a system of "carbon dioxide permits" to mitigate global warming. The Royal Society criticized the climate-change levy as an "inefficient way of reducing carbon-dioxide emissions because it does not apply to households and the transport sector. In addition, it penalizes electricity sources that do not produce greenhouse gases" (Houlder [November 18] 2002, 4).

Sir Eric Ash, who chaired the report's working group, urged the government to address the issue of charging for carbon dioxide emissions in its forthcoming White Paper on energy. "The U.K.'s carbon dioxide emissions have started to rise again. If the government's white paper misses the opportunity to reverse this trend, the U.K. could hasten the onset of potentially catastrophic climate change" (Houlder [November 18] 2002, 4). Their proposal would add an average of 15 percent to most British electricity bills and 10 percent to the cost of gasoline.

The Royal Society report suggested that a carbon tax might be equivalent initially to an extra one pence per kilowatt-hour for electricity bills or six pence more per liter of gasoline. The proposed tax should apply to all producers of carbon dioxide, including householders.

Vulnerable people could be compensated, for example, "through state pensions" (Houlder [November 18] 2002, 4).

The report dismissed concerns that a carbon tax or permit system would cause significant harm to the economy. It said the tax would produce substantial revenues that could be recycled, resulting in a modest cost to the overall economy. It acknowledged that the tax would affect some sectors, such an energy-intensive companies, but said that any measure to shield them should be time-limited (Houlder [November 18] 2002, 4). This report asserted that the introduction of a carbon tax would likely benefit wind, wave, tidal, nuclear energy, and carbon sequestration technology. By improving the viability of nonfossil fuel energy, the tax could ease reliance on imported energy and boost the security of energy supplies.

Another form of taxing system relies on private markets. In January 2004, the European Union imposed mandatory limits on industries' emissions of carbon dioxide and other greenhouse gases, introducing a market-based system for buying and selling the right to emit carbon. Companies that have met targets may thus sell "credits" to those that haven't.

WIND POWER CAPACITY SURGES

By the early twenty-first century, wind power was becoming competitive in cost with electricity generated by fossil fuels, as its use surged. Although wind power still was a tiny fraction of energy generated in the United States, some areas of Europe (Denmark, as well as parts of Germany and Spain) were using it as a major source.

The phrase "wind farm" entered the English language in the late 1990s, as the British began a concerted effort to harness the wind of its sea breezes for energy. Advances in wind-turbine technology adapted from the aerospace industry have reduced the cost of wind power from thirty-eight cents per kilowatt-hour (during the early 1980s) to three to six cents. This rate is competitive with the costs of power generation from fossil fuels, but costs vary according to site. Major corporations, including Shell International, have been moving into wind power.

Wind turbines. Courtesy of Getty Images/PhotoDisc.

The *London Guardian* in 2001 complained that Britain, once a leader in wind farm technology, gave up its development as Denmark accelerated its own work in the area. "While the nuclear and fossil-fuel lobbies, Thatcherite politicians, and the right-wing media were sneering at the crackpot idea of electricity being generated commercially from the wind, Denmark got on with making it happen," wrote the *Guardian* ("Power Games" 2001, 2). As a result, by the year 2001, more people (30,000) worked in the Danish wind-power industry than in Great Britain's coal mines. With its wind turbines producing electricity at four to five cents per kilowatt-hour, Denmark by 2004 was generating 20 percent of its electricity from wind power. Britain's production at the same time was 0.5 percent (Melvin 2004, 8-A). England later adopted wind technology to cut carbon emissions, but initially nearly all its turbines were imported from Denmark.

Wind-power generating capacity experienced record growth worldwide during 2002, with a 28 percent year-on-year increase. The European Wind Energy Association (EWEA) and the American Wind Energy

Association said that total capacity increased to 31,128 megawatts at the end of 2002 after a record-high 6,868 megawatts of new capacity was installed during the year. Capacity in European Union countries rose to 23,056 megawatts, or 74 percent of the world's total, reflecting the region's aggressive clean-energy policies. North America followed at 4,923 megawatts, with 3,150 megawatts in the rest of the world, including 415 megawatts in Japan ("World Wind-Power Capacity" 2003). Between 1995 and 2002, worldwide global wind-generating capacity increased nearly fivefold. During the same period, the use of coal, the main alternative for generating electricity and a major contribution to global warming, fell by 9 percent (P. Brown 2003, 13).

On a country-by-country basis, Germany topped the list in 2002 with a generating capacity of 12,001 megawatts, which supplied almost 5 percent of the country's total electricity demand. Schleswig-Holstein, the northernmost state of Germany, reached 20 percent wind-generated power by 2002; in some parts of that state, wind was generating 75 percent of demand. In Schleswig-Holstein, the wind energy industry has become the second-largest employer after tourism. More than 30,000 wind-power-related jobs had been created in Germany by 2001, "as private firms building rotors, towers, transfer stations and ever more powerful turbines have sprung up across wind-rich coastal states (Williams 2001, A-1).

By 2002, Spain generated 4,830 megawatts of wind power. Spain's industrial state of Navarra, which generated no wind power in 1996, by 2002 generated 25 percent of its electricity that way. By 2002, India had 1,702 megawatts of wind energy capacity. India had increased its use of wind power thirtyfold in ten years. Even with such rapid increases in capacity, wind power in 2002 accounted for only 0.4 percent of total electricity supply worldwide ("World Wind-Power Capacity" 2003).

The United States has been taking some tentative steps toward supplementing fossil fuels with wind energy and was generating 4,685 megawatts by 2002. Plans for Freedom Tower, which is being built in New York City on the site occupied by the World Trade Center's twin towers until September 11, 2001, will include the world's tallest wind farm, with as many as thirty wind turbines. Early in 2003, an electricity company partly owned by billionaire Warren Buffett announced plans to build the world's largest wind farm. The $323 million wind project will be sited in northwest or north-central Iowa. If approved by state agencies, the 310-megawatt project will be built by MidAmerican Energy Company. Based in Des Moines, Iowa, MidAmerican Energy is a unit of MidAmerican Energy Holdings Company, the majority of which is owned by Buffett's holding company, Berkshire Hathaway.

By 2004, the British government was requiring electricity-supply companies to increase their generation from renewable sources (the most prominent being wind power) to 10 percent by 2010 and 15 percent by 2015. Utilities that fail to meet these targets will become liable for fines to be paid, not to the government but, to other companies that have met the goals. Britain's government in mid-2003 announced plans to expand massively its reliance on wind power in generating electricity by opening up three offshore areas for the construction of wind farms, to include a string of wind farms built around the country's long and often blustery coastline. Energy from the planned new developments will eventually be enough to power more than 3.5 million households. The plans call for clusters of as many as 300 massive turbines five miles off the coasts of northwest England and north Wales, East Anglia in eastern England, and the estuary of the River Thames to the southeast of London. Britain's government has pledged to provide 10 percent of Britain's energy from renewable sources by 2010.

In another project, off the coast near Blyth, Northumberland, England, two giant windmills by the year 2000 were supplying electricity to 6,000 households. In an area known best for its fishing fleet, an annual music festival, and the distillation of whisky, Scotland's Kintyre Peninsula in mid-2002 became home to the United Kingdom's most efficient wind farm. The forty-six wind turbines on the peninsula's highest hill deliver thirty megawatts, enough to supply electricity to 25,000 homes. Scottish Power's £21 million (US $32.4 million) wind energy project is efficient because the Kintyre Peninsula is one of the windiest spots in Europe.

WIND POWER: THE DANISH EXAMPLE

The Danes have become world leaders in wind-power technology, establishing an industry that provides 150,000 Danish families with shares of profits from the national electricity grid. In matters of advanced technology, Denmark dominates the worldwide wind-power industry. Danish companies have supplied more than half the wind turbines now in use worldwide, making wind energy technology one of the country's largest exports. According to one account, "Wind turbines dot the Danish countryside like gigantic pinwheels, generating 20 percent of the country's electricity by 2003. The government plans to increase that figure to 50 percent by 2030. Thousands of rural Danish residents have joined wind-power cooperatives, buying turbines and leasing sites to build them, often on members' land" (Woodard 2001, 7). At the same time, power companies in Denmark are now taxed for each ton of carbon dioxide emitted above a low (and, over time, gradually declining) limit.

Peter Marsh, writing in London's *Financial Times*, described the Danish wind-power effort:

> Arriving in the Danish capital by sea, the visitor's first glimpse of the country is the Middelgrunden Wind Farm, a row of 20 enormous wind turbines gently spinning above the waves nearly two miles offshore. Completed in December [2001], Middelgrunden is the world's largest offshore wind farm, dominating the views from Copenhagen's docks and seaside parks. With blades 100 feet long, its wind machines generate enough power to supply 32,000 households. It's a fitting introduction to Denmark, a nation of 5 million that has emerged as the undisputed leader in wind energy. One Danish company, Vestas, which declared bankruptcy in 1986 as a maker of cranes and farm equipment, had reemerged by 2002 as the world's fourth-largest maker of wind turbines, with a market capitalization of U.S. $33.2 billion. (Marsh 2002, 30)

According to another on-site observer, "A graceful arc of alabaster windmills rises out of the midnight-blue waters of the Oresund strait, visual testimony to Danes' commitment to clean energy and the health of the planet" (Williams 2001, A-1).

The Danes are organizing local residents into wind-power investor co-ops, as Denmark corners the wind technology market worldwide at the same time it surpasses its own carbon dioxide reduction targets. By 2001, Denmark had the capacity to produce 2,364 megawatts, 450 watts per capita, in contrast with nine for each U.S. citizen. The Danes plan to generate 1,500 more megawatts per capita by 2010.

WIND POWER IN TEXAS OIL COUNTRY

Cielo Wind Power of Austin has been buying sixty-year "wind rights" from ranchers in Texas with plans to build the Noelke Hill Wind Ranch for $130 million. The wind ranch plans to generate 240 megawatts of power, enough to provide electricity to 80,000 homes, through TXU, a utility based in Dallas. Texas law now requires that 3 percent of the state's electricity come from renewable sources by 2009. Several companies, among them American Electric power and General Electric, have plans to spend as much as $1 billion on wind energy in the midst of the Permian Basin, which has heretofore been known as oil country. Wind speed in the area averages sixteen miles an hour and is consistent enough to bend juniper trees.

By late 2002, Cielo had built $300 million worth of wind-power turbines in Texas and sold them to various power companies, the turbines "jutting more than 200 feet into the air with blades 200 feet in diameter, to capture high-velocity wind patterns" (Herrick 2002, B-3). Texas by late 2002 was generating enough wind power to supply about 300,000 homes. A standard line used by wind developers with the ranchers in the area is, "You've been getting your hat blown off your head your whole life. It's time to stop cussin' and make some money" (p. B-3). Wind turbines placed roughly one per twenty-five acres earn the average landowner about $3,200 annually per turbine.

PHOTOVOLTAIC CELLS FOR SOLAR POWER

In southwestern Germany, Freiburg has become the world's first "solar city," with a "solar-powered train station, energy-efficient row houses, innumerable rooftop photovoltaic systems and, high on a hill overlooking the vineyards, the world-famous Heliotrop, a high-tech cylindrical house that rotates to follow the sun" (Roberts 2004, 192). Freiberg is home to the Fraunhofer Institute for Solar Energy Systems, where scientists have been seeking breakthrough research that will reduce the cost of photovoltaic solar energy to competitive levels with fossil fuels and nuclear power, including a "multilayer" P.V. cell with an efficiency of 40 percent, twice present levels. Germany and Japan have implemented tax subsidies for rooftop solar energy systems. Given trends in research and cost reduction, solar energy in regions with

Rows of solar-power panels. Courtesy of Corbis.

abundant sunshine (such as the Middle East, Mediterranean, and U.S. Southwest) may be cost-effective before the year 2010.

In England, Susan Roaf's solar roof fuels her electric car. Roaf, an architect at England's Oxford Brookes University, has designed a solar house fitted with photovoltaic cells that harness the sun's energy and convert it into electricity. According to a report in the *London Times*, "Her system is so efficient that she uses it to charge her electric car's batteries and makes a profit by selling 57 per cent back to the national network" (O'Connell 2002). Using similar technology, cities could someday become self-sufficient power generators without the use of fossil fuels.

"Once [solar] fuel cells have been perfected, we could all own one. If we were to convert our homes to solar power-using solar panels to heat the house, and P.V. cells to convert the Sun's energy into electricity, we could not only store the excess for a rainy day but could sell it back to the National Grid. This would be no bad thing," Roaf said. The English government has made £20 million available in the form of grants for conversion to solar energy. Currently photovoltaic cells are very expensive (Roaf's roof cost £25,000 in 1995), but the price will decrease as more people become energy self-sufficient. "After all, look at it this way," she said. "The roof is earning its keep" (O'Connell 2002).

Photovoltaic solar power is being sampled around the world. Despite its reticence about global warming, the George W. Bush administration in January 2003 allowed the National Park Service to install the first-ever solar electric system on the grounds of the White House. The Park Service, which manages the White House, installed a nine-kilowatt, rooftop solar electric photovoltaic system, as well as two solar thermal systems that are used to heat water.

In some northern Chinese farming villages, homes that a few years ago relied solely on low-grade coal for heat and cooking now utilize solar-heated rooftop water tanks (Landauer 2002, F-4). About 85 percent of Israeli households have solar water heaters, which the government estimates lighten the country's overall energy burden by 3 percent. Solar heaters have been used in Israel since the 1950s. The global energy crisis of 1979 reminded Israelis of their reliance on foreign sources for oil and coal. Solar water heaters have been required in new homes by law in Israel since 1980. About twenty companies, employing 4,000 people, manufacture the systems and sell them for $300 to $1,000 each. A mid-range system can pay for itself in energy savings in three or four years and, if maintained, can last more than ten years (Kaplow 2001, 1-P).

HYDROGEN FUEL–CELLED TRANSPORT: NO FREE LUNCH

Political correctness vis-à-vis global warming in the automobile in-dustry has become associated with development of hydrogen fuel cells, especially after President George W. Bush used his State of the Union address in January 2003 to propose $1.2 billion in research funding to develop hydrogen fuel technologies. With those funds, Bush said that America could lead the world in developing clean, hydrogen-powered automobiles. Iceland has, meanwhile, made plans to become the world's first hydrogen economy (utilizing its geothermal resources). Reykjavik's bus fleet has been retrofitted with fuel-cell engines, and hydrogen fueling stations have opened.

Jeremy Rifkin, a liberal social critic and author, wrote a book pub-lished September 2002 titled *The Hydrogen Economy: The Creation of the Worldwide Energy Web and the Redistribution of Power on Earth*. Rifkin believes that cheap hydrogen could make the twenty-first century more democratic and decentralized, much the way oil transformed the nine-teenth and twentieth centuries by fueling the rise of powerful corpora-tions and nation-states. With hydrogen, writes Rifkin, "Every human being on Earth could be 'empowered'" (Coy 2002, 83).

On September 5, 2002, a coalition of companies, including DuPont and 3M, asked Congress to spend $5.5 billion during the ensuing decade to advance fuel-cell development (Coy 2002, 83). Ford, during October 2002, announced that it would have a hydrogen car, the Focus, in limited service for commercial fleets in 2004. The car is an electric hybrid with a potential top speed of 115 miles an hour and was said to perform on a par with a gas-driven family car. In 2002, General Motors' Chief Executive Officer Richard Wagoner announced that company's launch of a $500 million hydrogen car for initial rollout in 2008. The cost could rise to billions of dollars annually if Wagoner chooses to mass-produce the new models. Wagoner defended the expense as a crucial investment in GM's future. "People say, 'How can you afford to spend so much on fuel cells?' and I say, 'How can you afford not to?'" (Lippert 2002, E-3). At the Paris Motor Show during September 2002, General Motors unveiled a Hy-wire, a "concept car" that uses a hydrogen fuel cell, about six years behind Toyota and Honda. Ford Motor has purchased a sizable stake in British Columbia's Ballard Power, a world leader in the race to provide the first commercially available hydrogen fuel–celled automobile. General Motors also owns a stake in this company.

Although it has been touted as pollution-free, hydrogen fuel is no free climatic lunch. Hydrogen, unlike oil or coal, does not exist in nature in a combustible form. Hydrogen is usually bonded with other chemical elements, and stripping them away to produce the pure hydrogen necessary to power a fuel cell requires large amounts of energy. Unless an alternative source (such as Iceland's geothermal resource) is available, hydrogen fuel usually is produced from fossil fuels. Extraction of hydrogen from water via electrolysis and compression of the hydrogen to fit inside a tank that can be used in an automobile requires a great deal of electricity. Until electricity is routinely produced via solar, wind, and other renewable sources, the hydrogen car will require energy from conventional sources, including fossil fuels. Today, 97 percent of the hydrogen produced in the United States comes from processes that involve the burning of fossil fuels, including oil, natural gas, and coal.

Paul M. Grant, writing in *Nature*, provided an illustration: "Let us assume that hydrogen is obtained by 'splitting' water with electricity—electrolysis. Although this isn't the cheapest industrial approach to 'make' hydrogen, it illustrates the tremendous production scale involved—about 400 gigawatts of continuously available electric power generation [would] have to be added to the grid, nearly doubling the present U.S. national average power capacity." That, calculated Grant, would represent the power-generating capacity of 200 Hoover Dams (Grant 2003, 129–130). At $1,000 per kilowatt, the cost of such new infrastructure would total about $400 billion. What about producing the 400 gigawatts with renewable energy? Grant estimated that "with the wind blowing hardest, and the sun shining brightest," wind-power generation would require a land area the size of New York State, or a layout of state-of-the-art photovoltaic solar cells half the size of Denmark (p. 130). Grant's preferred solution to this problem is to use energy generated by nuclear fission.

Ignoring hydrogen fuel cells' limitations, the European Union has advocated a transition to it from fossil fuels. The plan includes a $2 billion E.U. commitment over the course of several years to bring industry, the research community, and governments together to make this transition. According to Rifkin, "The E.U. decision to transform Europe into a hydrogen economy over the course of the next half century is likely to have as profound and far-reaching an impact on commerce and society as the changes that accompanied the harnessing of steam power and coal at the dawn of the industrial revolution and the introduction of the internal-combustion engine and the electrification of society in the 20th century" ("E.U. Plans" 2002).

THE HYDROGEN ECONOMY AND STRATOSPHERIC OZONE DEPLETION

Advocates of a hydrogen economy generally have ignored another potential problem. Some research indicates that leakage of hydrogen gas could cause problems in the Earth's stratospheric ozone layer. Writing in *Science*, researchers from the California Institute of Technology reported that the accumulation of leakage associated with a hydrogen economy could indirectly cause as much as a 10 percent decrease in stratospheric ozone.

Tracey K. Tromp and colleagues wrote: "The widespread use of hydrogen fuel cells could have heretofore unknown environmental impacts due to unintended emissions of molecular hydrogen, including an increase in the abundance of water vapor in the stratosphere (plausibly as much as about 1 part per million by volume). This would cause stratospheric cooling, enhancement of the heterogeneous chemistry that destroys ozone, an increase in noctilucent clouds, and changes in tropospheric chemistry and atmosphere-biosphere interactions" (Tromp et al. 2003, 1740).

If hydrogen replaced all fossil fuels for transportation and to power buildings, Tromp and colleagues estimated that 60–120 trillion grams of hydrogen would be released into the atmosphere each year, four to eight times the amount released today from human sources. The scientists assumed a 10–20 percent loss rate due to leakage. Molecular hydrogen freely rises and mixes with stratospheric air, resulting in creation of additional moisture at high altitudes, increased dampening of the stratosphere, and, through a chain of chemical reactions, cooling of the air and accelerating the destruction of ozone ("Hydrogen Leakage" 2003). The authors of this study estimated that hydrogen leakage could triple the amount of hydrogen in the stratosphere.

"We have an unprecedented opportunity this time to understand what we are getting into before we even switch to the new technology," said Tromp, the study's lead author. "It will not be like the case with the internal combustion engine, when we started learning the effects of carbon dioxide decades later" ("Hydrogen Leakage" 2003).

Refuting assertions that hydrogen leakage could deplete ozone, Martin G. Schultz and colleagues, writing in *Science*, asserted that "a possible rise in atmospheric hydrogen concentrations is unlikely to cause significant perturbations of the climate system" (Schultz et al. 2003, 624).

CREATION OF AN ORGANISM THAT WILL CONSUME CARBON DIOXIDE

J. Craig Venter, who compiled a human genetic map with private money, has decided to tackle the problem of global warming with a $100 million research endowment created from his stock holdings. Venter plans to scour deep ocean trenches for bacteria that can convert carbon dioxide to solid form with very little sunlight or other energy. If he cannot find such an organism in nature, Venter may build one himself. His idea is to devise technology that will allow humankind to

J. Craig Venter. Photo by Peter Cutts.

continue producing energy while lowering emissions of carbon dioxide. Ideally, Venter's organism or group of organisms will be able to take in carbon dioxide, break it down, and produce both biological compounds and energy (Gillis 2002, E-1).

Venter would like to invent two synthetic microorganisms. One would consume carbon dioxide and turn it into raw materials comprising the kinds of organic chemicals that are now made from oil and natural gas. The other microorganism would generate hydrogen fuel from water and sunshine. "Other groups are considering capturing carbon dioxide and pumping it down to the bottom of the ocean, which would be insane," Venter told the *London Financial Times*. "Why risk irreversible damage to the ocean when we could be doing something useful with the carbon?" (Cookson and Firn 2002, 11).

"We've barely scratched the surface of the microbial world out there to try to help the environment," Venter said. "We're going to be searching for some dramatic new microbes." Venter said that his ventures will be established as not-for-profit corporations. "I'm not in business anymore," he said (Gillis 2002, E-1). Venter, who calls his organization the Institute for Biological Energy Alternatives, will seek grant money from the U.S. Department of Energy. His goal will be to explore whether modern science can use the power of biology to solve the world's most serious environmental crisis.

Venter has proposed installing colonies of organisms in "bioreactors" near power plants to consume emissions of carbon dioxide and turn the gas into solids such as sugars, proteins, and starches, mimicking the behavior of green plants. According to an account in the *Washington Post*, "Venter plans to base his approach on one of the most striking developments in biology in recent years—the discovery, in deep ocean trenches and volcanic hot spots on the ocean floor, of a wide array of bacteria that can perform extensive chemical reactions without needing sunlight. These are thought to be descendants of the most primitive life forms that arose on the Earth, and scientists are just beginning to explore their potential" (Gillis 2002, E-1).

A POLYMER THAT ABSORBS CARBON DIOXIDE AT HIGH TEMPERATURES

Early in 2002, scientists revealed details of an inexpensive technology that captures carbon dioxide from industrial processes. According to a report by the Environment News Service, scientists at the Department of Energy's Los Alamos National Laboratory (LANL) "are developing a new high-temperature polymer membrane to separate and capture carbon dioxide, preventing its escape into the atmosphere. This work is part of the Department of Energy's Carbon Sequestration Program, which is exploring ways to capture carbon dioxide from fossil fuel burning and reduce human impacts on climate" ("Hot Polymer" 2002).

This technology is aimed at sequestering the 30 percent of anthropogenic carbon dioxide that comes from power-producing industries. Present technology is limited to dealing with waste carbon dioxide emitted by such plants up to 150 degrees C; the wastes often reach 375 degrees C. Speaking at an American Geophysical Union conference May 29, 2002, in Washington, D.C., Jennifer Young, principal investigator for LANL's carbon dioxide membrane separation project, described a new polymeric-metallic membrane that is stable at temperatures as high as 370 degrees C. "Current technologies for separating carbon dioxide from other gases require that the gas stream be cooled to below 150 degrees Celsius, which reduces energy efficiency and increases the cost of separation and capture," said Young. "By making a membrane which functions at elevated temperatures, we increase the practicality and economic feasibility of using membranes in industrial settings" ("Hot Polymer" 2002).

TOKYO'S SEAWATER COOLING GRID

Tokyo city planners during 2002 announced plans to build the world's largest cooling system and fill it with seawater to reduce temperatures by 2.6 degrees C—or about the amount that average temperatures in the city increased during the twentieth century. General warming is being enhanced in Tokyo by an intensifying urban heat-island effect as population density increases and more air conditioning (with its waste heat) is being used to cool tall buildings. This huge construction project, estimated to cost $300 billion, would require "a lattice of pipes under 250 hectares (about 600 acres) of the city, drawing in cold water from Tokyo Bay. Heat would be forced into the network through a second set of pipes connected to air conditioning systems in the buildings above. The heat would be absorbed by the seawater, which would then flow back into the bay" (Ryall 2002, 20). Skeptics of the project fear that dumping warmed seawater into Tokyo Bay could cause environmental damage.

Tokyo's average temperature rise of 2.9 degrees C during the twentieth century was five times the average worldwide rate of warming. The number of days when the city experiences temperatures of 30 degrees C or more has doubled during the last two decades. "This is a radical approach but that's what the situation demands," said Tadafumi Maejima, director of the Japan District Heating and Cooling Association, which was commissioned to draw up the plan. "Nothing like this has been attempted anywhere in the world, but we could be ready to start work in as little as two years" (Ryall 2002, 20).

According to a report in the *London Times*, "The pipes would initially spread beneath 123 hectares of the Marunouchi district, the city's financial heart. Later they would extend through Kasumigaseki, where the Diet [parliament], and the ministries are based. A further phase, covering an area north of Tokyo station, would use water from a sewage treatment plant.... The amount of heat emitted by buildings needs to be cut, especially in the heart of the city where demand for air conditioners is concentrated, to break the vicious cycle of summer warming" (Ryall 2002, 20).

ETHANOL TRUMPS GASOLINE IN BRAZIL

The idea of using carbohydrates as fuel is not new; Henry Ford's first model was designed to run on it; during World War II, the fuel was used to stretch supplies of gasoline. After thirty years of intense effort, Brazil by 2005 had developed an ethanol market, mainly using sugarcane, that had replaced 20 percent of its gasoline usage. During the 1970s, the country's government began developing the ethanol industry by subsidizing the sugarcane industry and requiring its use in government vehicles, By the late 1990s, however, the subsidies were phased out as the cost of producing ethanol dropped to 80 cents a gallon, less than $1.50, the world-wide average for producing gasoline (Luhnow and Samor 2006, A-1, A-8).

Brazil's use of gasoline has declined slightly since the 1970s. Seven of every ten new cars in Brazil are "flex-fuel," fitted to consume gasoline or ethanol, or a mixture of both. Sugarcane is the least expensive plant mass for ethanol, and Brazil has an abundance of it. Brazil exported $600 million worth of ethanol in 2005. Brazil's government taxes ethanol at 9 cents per gallon, compared to 42 cents for gasoline, and all gasoline is legally required to contain at least 10 percent ethanol. Researchers in Brazil have decoded the genetics of sugarcane and used the knowledge to breed varieties with higher sugar content. Brazil has increased the per-acre productivity of sugarcane threefold since 1975 (Luhnow and Samor 2006, A-1, A-8).

REFERENCES:
PART VI.
SOLUTIONS

"Australian Scientists Announce Good News at Last on Global Warming." Agence France Presse, November 25, 2003. (Lexis).

Betts, Richard A. "Offset of the Potential Carbon Sink from Boreal Forestation by Decreases in Surface Albedo." *Nature* 408 (November 9, 2000): 187–190.

Boyd, Philip W., Andrew J. Watson, Cliff S. Law, Edward R. Abraham, Thomas Trull, Rob Murdoch, et al. "A Mesoscale Phytonplankton Bloom in the Polar Southern Ocean Stimulated by Iron Fertilization." *Nature* 407 (October 12, 2000): 695–702.

Bradsher, Keith. "Ford Tries to Burnish Image by Looking to Cut Emissions." *New York Times*, May 4, 2001, C-3.

Brown, Amanda. "New Figures Show Fall in Greenhouse Gas Emissions." Press Association (United Kingdom), March 27, 2003. (Lexis).

Brown, Paul. "Analysis: Blair Sets Out Far-Reaching Vision but Where Are the Practical Policies?" *London Guardian*, February 25, 2003, 13.

Buesseler, Ken O., and Philip W. Boyd. "Will Ocean Fertilization Work?" *Science* 300 (April 4, 2003): 67–68.

Carey, John, and Sarah R. Shapiro. "Consensus Is Growing among Scientists, Governments, and Business That They Must Act Fast to Combat Climate Change. This Has Already Sparked Efforts to Limit CO_2 Emissions. Many Companies Are Now Preparing for a Carbon-Constrained World." *Business Week*, August 16, 2004. (Lexis).

Catholic Bishops, U.S. Conference. "Global Climate Change: A Plea for Dialogue, Prudence, and the Common Good: A Statement of the U.S. Catholic Bishops." Ed. William P. Fay, June 15, 2001. www.ncrlc.com/climideas. html.

Chisholm, Sallie W. "Stirring Times in the Southern Ocean." *Nature* 407 (October 12, 2000): 685–686.

Clayton, Chris. "U.S.D.A. Will Offer Incentives for Conserving Carbon in Soil." *Omaha World-Herald*, June 7, 2003, D-1, D-2.

Clover, Charles, and David Millward. "Future of Cheap Flights in Doubt: Ban New Runways and Raise Fares, Say Pollution Experts." *London Daily Telegraph*, November 30, 2002, 1, 4.

"Conferences Tackle Key Issues in Air Conditioning, Refrigeration." AScribe Newswire, June 23, 2004. (Lexis).

Cookson, Craig, and David Firn. "Breeding Bugs That May Help Save the World: Craig Venter Has Found a Large Project to Follow the Human Genome." *London Financial Times*, September 28, 2002, 11.

Coy, Peter. "The Hydrogen Balm? Author Jeremy Rifkin Sees a Better, Post-Petroleum World." *Business Week*, September 30, 2002, 83.

Daynes, Byron W., and Glen Sussman. "The 'Greenless' Response to Global Warming." *Current History*, December 2005, 438–443.

Dunne, Nancy. "Climate Change Research Sparks Hawaii Protests." *London Financial Times*, June 20, 2002, 2.

Eggert, David. "Shareholders File Global Warming Resolutions at Ford, GM." Associated Press, December 11, 2002. (Lexis).

"E.U. Plans to Become First Hydrogen Economy Superpower." *Industrial Environment* 12 (13) (December 2002). (Lexis).

Gillis, Justin. "A New Outlet for Venter's Energy: Genome Maverick to Take On Global Warming." *Washington Post*, April 30, 2002, E-1.

"Global Warming Could Hamper Ocean Sequestration." Environment News Service, December 4, 2002. http://ens-news.com/ens/dec2002/2002-12-04-09.asp.

"Global Warning on Climate." *Montreal Gazette*, February 27, 2001, B-2.

Grant, Paul M. "Hydrogen Lifts Off—with a Heavy Load: The Dream of Clean, Usable Energy Needs to Reflect Practical Reality." *Nature* 424 (July 10, 2003): 129–130.

Hakim, Danny. "Ford Executives Adopt Ambitious Plan to Rein in Global Warming." *New York Times* in *International Herald-Tribune*, October 5, 2004, 11.

———. "Ford Stresses Business, but Disappoints Environmentalists." *New York Times*, August 20, 2002, C-4.

———. "Several States Likely to Follow California on Car Emissions." *New York Times*, June 11, 2004, C-4.

Hansen, James E., and Makiko Sato. "Trends of Measured Climate Forcing Agents." *Proceedings of the National Academy of Sciences* 98 (December 18, 2001): 14778–14783.

Harper's Index, August 2005, 11.

Heilprin, John. "Study Sees Economic Benefits of Reducing Global Warming." Associated Press, July 13, 2001. www.worldwildlife.org/climate/climate.cfm.

Herrick, Thaddeus. "The New Texas Wind Rush: Oil Patch Turns to Turbines, as Ranchers Sell Wind Rights: A New Type of Prospector." *Wall Street Journal*, September 23, 2002, B-1, B-3.

Hoffert, Martin I., Ken Caldeira, Gregory Benford, David R. Criswell, Christopher Green, Howard Herzog, et al. "Advanced Technology Paths to Global Climate Stability: Energy for a Greenhouse Planet." *Science* 298 (November 1, 2002): 981–987.

Hoffman, Ian. "Iron Curtain over Global Warming: Ocean Experiment Suggests Phytoplankton May Cool Climate." *Hayward (California) Daily Review*, April 17, 2004. (Lexis).

Holly, Chris. "World CO_2 Emissions Up 13 Percent from 1990–2000." *Energy Daily* 30 (206) (October 25, 2002). (Lexis).

"Hot Polymer Catches Carbon Dioxide." Environment News Service, May 29, 2002. http://ens-news.com/ens/may2002/2002-05-29-05.asp.

Houlder, Vanessa. "Rise Predicted in Aviation Carbon Dioxide Emissions." *London Financial Times*, December 16, 2002, 2.

———. "Royal Society Calls for Carbon Levy or Permits." *London Financial Times*, November 18, 2002, 4.

"Hydrogen Leakage Could Expand Ozone Depletion." Environment News Service, June 13, 2003. http://ens-news.com/ens/jun2003/2003-06-13-09.asp.

Ingham, John, and Barry Keevins. "U.K. Set to Be 6 C. Warmer by 2099; London May Be Just Like Madrid; Threat of Tornadoes; Cities Face Flooding; The Three-Day Winter." *London Express*, February 24, 2003, 17.

"Iron Link to CO_2 Reductions Weakened." Environment News Service, April 10, 2003. http://ens-news.com/ens/apr2003/2003-04-10-09.asp#anchor8.

John Paul II. *On the Hundredth Anniversary of Rerum Novarum (Centesimus Annus).* No. 32. Washington, D.C.: United States Catholic Conference, 1991.

Jones, George, and Charles Clover. "Blair Warns of Climate Catastrophe: 'Shocked' Prime Minister Puts Pressure on U.S. and Russia over Emissions." *London Daily Telegraph*, September 15, 2004, 2.

Kaplow, Larry. "Solar Water Heaters: Israel Sets Standard for Energy; Cutting Dependence: Jerusalem's Alternative Energy Use a Lesson for United States." *Atlanta Journal and Constitution*, August 5, 2001, 1-P.

Kondratyev, Kirill, Vladimir F. Krapivin, and Costas A. Varotsos. *Global Carbon Cycle and Climate Change*. Berlin, Germany: Springer/Praxis, 2003.

Kristof, Nicholas. "The Storm Next Time." *New York Times*, September 1, 2005. www.nytimes.com.

Lai, R. "Soil Carbon Sequestration Impacts on Global Climate Change and Food Security." *Science* 304 (June 11, 2004): 1623–1627.

Landauer, Robert. "Big Changes in Our China Suburb." *Sunday Oregonian*, October 20, 2002, F-4.

Lean, Geoffrey. "We Regret to Inform You That the Flight to Malaga Is Destroying the Planet: Air Travel Is Fast Becoming One of the Biggest Causes of Global Warming." *London Independent*, August 26, 2001, 23.

Leggett, Jeremy. *The Carbon War: Global Warming and the End of the Oil Era.* New York: Routledge, 2001.

"Limiting Methane, Soot Could Quickly Curb Global Warming." Environment News Service, January 16, 2002. http://ens-news.com/ens/jan2002/2002L-01-16-01.html.

Lippert, John. "G[eneral] M[otors] Chief Weighs Future of Fuel Cells." *Toronto Star*, September 27, 2002, E-3.

"Liquid CO_2 Dump in Norwegian Sea Called Illegal." Environment News Service, July 11, 2002. http://ens-news.com/ens/jul2002/2002-07-11-02.asp.

Luhnow, David, and Geraldo Samor. "As Brazil Fills Up on Ethanol, It Weans Off Energy Imports." *Wall Street Journal*, January 9, 2006, A-1, A-8.

Macalister, Terry. "Confused Esso Tries to Silence Green Critics." *London Guardian*, June 25, 2002, 21.

———. "Shell Chief Delivers Global Warming Warning to Bush in His Own Back Yard." *London Guardian*, March 12, 2003, 19.

"Maine Sets Global Warming Reduction Goals." Environment News Service, June 26, 2003. http://ens-news.com/ens/jun2003/2003-06-26-09.asp#anchor4.

"Many U.S. Industry Giants Ignoring Global Warming." Environment News Service, July 9, 2003. http://ens-news.com/ens/jul2003/2003-07-09-11.asp.

Marsh, Peter. "Progress Is a Long and Windy Road: A Once-Bankrupt Danish Crane Parts Maker Is Now the World's Biggest Maker of Wind Turbines." *London Financial Times*, May 30, 2002, 30.

Masson, Gordon. "Eco-Friendly Movement Growing in Music Biz." *Billboard*, March 15, 2003, 1.

McCarthy, Michael. "Ford Predicts End of Car Pollution: Boss Predicts the End of Petrol." *London Independent*, October 6, 2000, 10.

———. "World's Largest Windmills Start Contributing to United Kingdom National Grid." *London Independent*, December 8, 2000, 9.

McKay, Paul. "Ford Leads Big Three in Green Makeover: Henry Ford Brought the World the Internal Combustion Engine. His Great-Grandson May Consign It to History—and Help Preserve the Planet in the Bargain." *Ottawa Citizen*, May 28, 2001, A-1.

Melvin, Don. "Storm over Wind Energy: Britain's Renewable Power Push Stirs Turbulent Debate." *Atlanta Journal-Constitution*, July 5, 2004, 8-A.

"New E.U. Law Aims to Double Green Energy by 2010." Reuters, July 5, 2001. (Lexis).

"Norway Says No to Controversial Plan to Store CO_2 on Ocean Floor." Agence France Presse, August 22, 2002. (Lexis).

O'Connell, Sanjida. "Power to the People." *London Times*, May 20, 2002. (Lexis).

"One in Five ExxonMobil Shareholders Want Climate Action." Environment News Service, May 28, 2003. http://ens-news.com/ens/may2003/2003-05-28-09.asp#anchor3.

Pacala, S., and R. Socolow. "Stabilization Wedges: Solving the Climate Problem for the Next 50 Years with Current Technologies." *Science* 305 (August 13, 2004): 968–972.

Pearce, Fred. "Ground-Breaking Solutions to Global Warming." *London Independent*, December 8, 2000, 8.

Pianin, Eric. "On Global Warming, States Act Locally: At Odds with Bush's Rejection of Mandatory Cuts, Governors and Legislatures Enact Curbs on Greenhouse Gases." *Washington Post*, November 11, 2002, A-3.

Polakovic, Gary. "States Taking the Initiative to Fight Global Warming: Unhappy with Bush's Policies, Local Officials Work to Slow Climate Change." *Los Angeles Times*, October 7, 2001, A-1.

"Power Games: Britain Was a World Leader in Wind Farm Technology. So Why Are All Our Windmills Made in Denmark?" *London Guardian*, July 16, 2001, 2.

Renewing the Earth: An Invitation to Reflection and Action on Environment in Light of Catholic Social Teaching. Washington, D.C.: United States Catholic Conference, n.d., 3.

Revkin, Andrew C. "Antarctic Test Raises Hope on a Global-Warming Gas." *New York Times*, October 12, 2000, A-18.

———. "Climate Talks Shift Focus to How to Deal with Changes." *New York Times*, November 3, 2002, A-8.

———. "Planting New Forests Can't Match Saving Old Ones in Cutting Greenhouse Gases, Study Finds." *New York Times*, September 22, 2000, A-23.

———. "Scientists Say a Quest for Clean Energy Must Begin Now." *New York Times*, November 1, 2002, A-6.

"Rich Countries' Greenhouse Gas Emissions Ballooning." Environment News Service, June 9, 2003. http://ens-news.com/ens/jun2003/2003-06-09-02.asp.

Roberts, Paul. *The End of Oil: The Edge of a Perilous New World*. Boston: Houghton-Mifflin, 2004.

Royse, David. "Scientists: Bush Global Warming Stance Invites Stronger Storms." Associated Press, October 25, 2004. (Lexis).

Ryall, Julian. "Tokyo Plans City Coolers to Beat Heat." *London Times*, August 11, 2002, 20.

Schiermeier, Quirin. "The Oresmen." *Nature* 421 (January 9, 2003): 109–110.

Schrope, Mark. "Global Warming: A Change of Climate for Big Oil." *Nature* 411 (May 31, 2001): 516–518.

Schultz, Martin G., Thomas Diehl, Guy P. Brasseur, and Werner Zittel. "Air Pollution and Climate-Forcing Impacts of a Global Hydrogen Economy." *Science* 302 (October 24, 2003): 624–627.

Schulze, Ernst-Detlef, Christian Wirth, and Martin Heimann. "Managing Forests after Kyoto." *Science* 289 (September 22, 2000): 2058–2059.

Seelye, Katharine. "Environmental Groups Gain as Companies Vote on Issues." *New York Times*, May 29, 2003, C-1.

Seibel, Brad A., and Patrick J. Walsh. "Potential Impacts of CO_2 Injection on Deep-Sea Biota." *Science* 294 (October 12, 2001): 319–320.

Shaw, M. Rebecca, Erika S. Zavaleta, Nona R. Chiariello, Elsa E. Cleland, Harold A. Mooney, and Christopher B. Field. "Grassland Responses to Global Environmental Changes Suppressed by Elevated CO_2." *Science* 298 (December 6, 2002): 1987–1990.

"Ski Resorts Get Creative to Battle Global Warming." Environment News Service, February 20, 2003. http://ens-news.com/ens/feb2003/2003-02-20-02.asp.

Speth, James Gustave. *Red Sky at Morning: America and the Crisis of the Global Environment*. New Haven, CT: Yale University Press, 2004.

Stokstad, Erik. "California Sets Goals for Cutting Greenhouse Gases." *Science* 308 (June 10, 2005): 1530.

Thiessen, Mark. "Researchers Fighting 'Bad' Breath in Cattle." Associated Press, June 8, 2003. (Lexis).

Tromp, Tracey K., Run-Lie Shia, Mark Allen, John M. Eiler, and Y. L. Yung. "Potential Environmental Impact of a Hydrogen Economy on the Stratosphere." *Science* 300 (June 13, 2003): 1740–1742.

Tsuda, Atsushi, Shigenobu Takeda, Hiroaki Saito, Jun Nishioka, Yukihiro Nojiri, Isao Kudo, et al. "A Mesoscale Iron Enrichment in the Western Subarctic Pacific Induces a Large Centric Diatom Bloom." *Science* 300 (May 9, 2003): 958–961.

Urquhart, Frank, and Jim Gilchrist. "Air Travel to Blame as Well." *The Scotsman*, October 8, 2002. (Lexis).

"U.S. Electricity Sector Makes Twice as Much Greenhouse Gas as Europe: Report." Agence France Presse, October 21, 2002. (Lexis).

Vidal, John, and Terry Macalister. "Kyoto Protests Disrupt Oil Trading." *London Guardian*, February 17, 2005, 4.

Watson, A. J., D. C. E. Bakker, A. J. Ridgwell, P. W. Boyd, and C. S. Law. "Effect of Iron Supply on Southern Ocean CO_2 Uptake and Implications for Glacial Atmospheric CO_2." *Nature* 407 (October 12, 2000): 730–733.

Watt, Nicholas. "Planet Is Running Out of Time, Says Meacher: U.S. Rejection of Kyoto Climate Plan 'Risks Uninhabitable Earth.'" *London Guardian*, May 16, 2002, 11.

Webster, Ben. "Boeing Admits Its New Aircraft Will Guzzle Fuel." *London Times*, June 19, 2001. (Lexis).

Williams, Carol J. "Danes See a Breezy Solution: Denmark Has Become a Leader in Turning Offshore Windmills into Clean, Profitable Sources of Energy as Europe Races to Meet Emissions Goals." *Los Angeles Times*, June 25, 2001, A-1.

"Wind Power Use Grows by 30 Per Cent." *London Guardian*, January 10, 2002, 15.

Winestock, Geoff. "How to Cut Emissions? E.U. Can't Decide." *Wall Street Journal*, July 13, 2001, A-9.

Wolf, Martin. "Hot Air about Global Warming." *London Financial Times*, November 29, 2000, 27.

Woodard, Colin. "Wind Turbines Sprout from Europe to U.S." *Christian Science Monitor*, March 14, 2001, 7.

"World Wind-Power Capacity Marks Record Growth for 2002." Japan Economic Newswire, March 3, 2003. (Lexis).

SELECTED BIBLIOGRAPHY

Abraham, Spencer. "The Bush Administration's Approach to Climate Change." *Science* 305 (July 30, 2004): 616–617.

Abrahmson, Dean Edwin. *The Challenge of Global Warming.* Washington, D.C.: Island Press, 1989.

Abram, Nerilie J., Michael K. Gagan, Malcolm T. McCulloch, John Chappell, and Wahyoe S. Hantoro. "Coral Reef Death during the 1997 Indian Ocean Dipole Linked to Indonesian Wildfires." *Science* 301 (August 15, 2003): 952–955.

"Acid Rain Emissions Halve in 11 Years." Hermes Database (Great Britain), May 20, 2003. (Lexis).

Adam, David. "Hatchoooooh! Record Numbers of People Are Complaining of Hay Fever." *London Guardian*, June 18, 2003, 4.

Adams, J. Brad, Michael E. Mann, and Casper M. Amman. "Proxy Evidence for an El Niño-Like Response to Volcanic Forcing." *Nature* 426 (November 20, 2003): 274–278.

Ainley, David G. *The Adelie Penguin: Bellweather of Climate Change.* New York: Columbia University Press, 2002.

———, G. Ballard, S. D. Emslie, W. R. Fraser, P. R. Wilson, E. J. Woehler, et al. "Adélie Penguins and Environmental Change." Letter to the Editor. *Science* 300 (April 18, 2003): 429.

Aitken, Mike. "St. Andrews Stymied by Natural Hazard." *The Scotsman*, April 18, 2001, 20.

"Alaskan Glaciers Retreating." Environment News Service, December 11, 2001. http://ens-news.com/ens/dec2001/2001L-12-11-09.html.

Alexander, Vera. "Arctic Marine Ecosystems." In *Global Warming and Biological Diversity*, ed. Robert L. Peters and Thomas E. Lovejoy, 221–232. New Haven, CT: Yale University Press, 1992.

Allakhverdov, Andrey, and Vladimir Pokrovsky. "Russia, Reluctantly, Backs Kyoto." *Science* 306 (October 8, 2004): 209.

Allen, Myles. "Film: Making Heavy Weather." *Nature* 429 (June 7, 2004): 347–348.

———, Sarah Raper, and John Mitchell. "Uncertainty in the I.P.C.C.'s Third Assessment Report." *Science* 293 (July 20, 2001): 430–433.

Allen, Myles R. "Climate Forecasting: Possible or Probable." *Nature* 425 (September 18, 2003): 242.

———, and William J. Ingram. "Constraints on Future Changes in Climate and the Hydrologic Cycle." *Nature* 419 (September 12, 2002): 224–232.

———, Peter A. Stott, John F. B. Mitchell, Reiner Schnur, and Thomas L. Delworth. "Quantifying the Uncertainty in Forecasts of Anthropogenic Climate Change." *Nature* 407 (October 5, 2000): 617–620.

Alley, Richard B., ed. *Abrupt Climate Change: Inevitable Surprises*. Committee on Abrupt Climate Change, Ocean Studies Board, Polar Research Board, Board on Atmospheric Sciences and Climate, Division of Earth and Life Sciences, National Research Council. Washington, D.C.: National Academy Press, 2002.

———. "Ice-Core Evidence of Abrupt Climate Changes." *Proceedings of the National Academy of Sciences* 97 (4) (February 15, 2000): 1331–1334.

———. "On Thickening Ice?" *Science* 295 (January 18, 2002): 451–452.

———. *The Two-Mile Time Machine: Ice Cores, Abrupt Climate Change, and Our Future*. Princeton, NJ: Princeton University Press, 2000.

———, Peter U. Clark, Philippe Huybrechts, and Ian Joughin. "Ice-Sheet and Sea-Level Changes." *Science* 310 (October 21, 2005): 456–460.

———, D. E. Lawson, G. J. Larson, E. B. Evenson, and G. S. Baker. "Stabilizing Feedbacks in Glacier-Bed Erosion." *Nature* 424 (August 14, 2003): 758–760.

———, J. Marotzke, W. D. Nordhaus, J. T. Overpeck, D. M. Peteet, R. A. Pielke Jr., et al. "Abrupt Climate Change." *Science* 299 (March 28, 2003): 2005–2010.

Alleyne, Richard, and Ben Fenton. "Heatwave Britain—When the Trees Turn Toxic." *London Daily Telegraph*, May 10, 2004, 3.

Alverson, Keith, Ray Bradley, Keith Briffa, Julia Cole, Malcolm Hughes, Isabelle Larocque, et al. "A Global Paleoclimate Observing System." *Science* 293 (July 6, 2001): 47.

———, Raymond S. Bradley, and Thomas F. Pedersen, eds. *Paleoclimate, Global Change, and the Future*. Berlin: Springer Verlag, 2003.

"Amazon Deforestation Causing Global Warming, Brazilian Government Says." British Broadcasting Corporation International reports, December 10, 2004. (Lexis).

Amor, Adlai. "Report Warns of Growing Destruction of World's Coastal Areas." World Resources Institute, April 17, 2001. www.dooleyonline.net/media_preview/index.cfm.

Anderson, David M., Jonathan T. Overpeck, and Anil K. Gupta. "Increase in the Asian Southwest Monsoon during the Past Four Centuries." *Science* 297 (July 26, 2002): 596–599.

Anderson, J. G., W. H. Brune, and M. H. Proffitt. "Ozone Destruction by Chlorine Radicals within the Antarctic Vortex: The Spatial and Temporal Evolution of ClO/O3, Anticorrelation Based on In Situ ER-2 Data." *Journal of Geophysical Research* 94 (1989): 11465–11479.

Anderson, J. W. "The History of Climate Change as a Political Issue." The Weathervane: A Global Forum on Climate Policy Presented by Resources for the Future, August, 1999. www.weathervane.rff.org/features/feature005.html.

Anderson, Theodore L., Robert J. Charlson, Stephen E. Schwartz, Reto Knutti, Olivier Boucher, Henning Rodhe, et al. "Climate Forcing by Aerosols—a Hazy Picture." *Science* 300 (May 16, 2003): 1103–1104.

Andreae, Meinrat O. "The Dark Side of Aerosols." *Nature* 409 (February 8, 2001): 671–672.

———, D. Rosenfeld, P. Artaxo, A. A. Casta, G. P. Frank, K. M. Longo, et al. "Smoking Rain Clouds over the Amazon." *Science* 303 (February 27, 2004): 1337–1342.

Andrews, David G. *An Introduction to Atmospheric Physics.* Cambridge, U.K.: Cambridge University Press, 2000.

"Antarctic Sea Ice Has Increased." Environment News Service, August 23, 2002. http://ens-news.com/ens/aug2002/2002-08-23-09.asp#anchor5.

Archer, David. "Ocean Science: Who Threw That Snowball?" *Science* 302 (October 31, 2003): 791–792.

Arendt, Anthony A., Keith A. Echelmeyer, William D. Harrison, Craig S. Lingle, and Virginia B. Valentine. "Rapid Wastage of Alaska Glaciers and Their Contribution to Rising Sea Level." *Science* 297 (July 19, 2002): 382–386.

"Are White Christmases Just a Memory?" Environment News Service, December 21, 2001. http://ens-news.com/ens/dec2001/2001L-12-21-09.html.

Arnold, David. "Global Warming Lends Power to Jellyfish in Narragansett Bay and Long Island Sound: Non-Native Species Are Taking Over." *Boston Globe,* July 2, 2002, C-1.

Aron, Joan L., and Jonathan A. Patz, eds. *Ecosystem Change and Public Health: A Global Perspective.* Baltimore: Johns Hopkins University Press, 2001.

Arrhenius, Svante. "On the Influence of Carbonic Acid in the Air upon the Temperature of the Ground." *The London, Edinburgh, and Dublin Philosophical Magazine and Journal of Science,* 5th ser., April 1896, 237–276.

Arthur, Charles. "Snows of Kilimanjaro Will Disappear by 2020, Threatening World-Wide Drought." *London Independent*, October 18, 2002, 7.

———. "Super El Niño Could Turn Amazon into Dustbowl: British Association for the Advancement of Science." *London Independent*, September 9, 2003, 6.

———. "Temperature Rise Kills 90 Per Cent of Ocean's Surface Coral." *London Independent*, September 18, 2003. (Lexis).

Artrill, M. J., and M. Power. "Climatic Influence on a Marine Fish Assemblage." *Nature* 417 (May 16, 2002): 275–278.

Asner, Gregory P., David E. Knapp, Eben N. Broadbent, Paulo J. C. Oliveira, Michaael Keller, and Jose N. Silva. "Selective Logging in the Brazilian Amazon." *Science* 310 (October 21, 2005): 480–481.

Assel, Raymond A., Frank H. Quinn, and Cynthia E. Sellinger. "Hydroclimatic Factors of the Recent Record Drop in Laurentian Great Lakes Water Levels." *Bulletin of the American Meteorological Society* 85 (8) (August 2004): 1143–1150.

Associated Press. "Pollution Adds to Global Warming." Via Excite! Data feed, 3:23 AM ET October 26, 2000. http://apple.excite.com/.

Association of American Geographers Global Change and Local Places Research Team. *Global Change and Local Places: Estimating, Understanding, and Reducing Greenhouse Gases*. New York: Cambridge University Press, 2003.

Atkinson, Angus, Volker Siegel, Evgeny Pakhomov, and Peter Rothery. "Long-Term Decline in Krill Stock and Increase in Salps within the Southern Ocean." *Nature* 432 (November 4, 2004): 100–103.

"Atmospheric Science: Really High Clouds." *Science* 292 (April 13, 2001): 171.

Austin, J., N. Butchart, and K. P. Shine. "Possibility of an Arctic Ozone Hole in a Doubled-CO_2 Climate." *Nature* 360 (November 19, 2001): 221–225.

Austin, Jay E., and Carl E. Bruch, eds. *The Environmental Consequences of War: Legal, Economic, and Scientific Perspectives*. Cambridge: Cambridge University Press, 2000.

"Australia Assesses Fire Damage in Capital." Associated Press in *Omaha World-Herald*, January 20, 2003, A-4.

"Australian Scientists Announce Good News at Last on Global Warming." Agence France Presse, November 25, 2003. (Lexis).

Ayres, Ed. *God's Last Offer: Negotiating a Sustainable Future*. New York: Four Walls Eight Windows, 1999.

Bagla, Pallava. "Climate Forecasting: Drought Exposes Cracks in India's Monsoon Model." *Science* 297 (August 23, 2002): 1265–1267.

Baily, Ronald, ed. *Global Warming and Other Myths: How the Environmental Movement Uses False Science to Scare Us to Death*. Roiseville, CA: Competitive Enterprise Institute, 2002.

Baines, Santo, Richard D. Norris, Richard M. Corfield, and Kristina L. Faul. "Termination of Global Warmth at the Palaeocene/Eocene Boundary through Productivity Feedback." *Nature* 407 (September 14, 2000): 171–173.

Baker, Andrew C. "Reef Corals Bleach to Survive Change." *Nature* 411 (June 14, 2001): 765–766.

———, Craig J. Starger, Tim R. McClanahan, and Peter W. Glynn. "Corals' Adaptive Response to Climate Change." *Nature* 430 (August 12, 2004): 741.

Baker, Linda. "The Hole in the Sky: Think the Ozone Layer Is Yesterday's Issue? Think Again." *E: The Environmental Magazine*, November/December 2000, 34–39. www.e-magazine.com/november-december 2000/1100feat2.html.

Baker, Paul A. "Trans-Atlantic Climate Connections." *Science* 297 (April 5, 2002): 67–68.

———, Geoffrey O. Seltzer, Sherilyn C. Fritz, Robert B. Dunbar, Matthew J. Grove, Pedro M. Tapia, et al. "The History of South American Tropical Precipitation for the Past 25,000 Years." *Science* 291 (January 26, 2001): 640–643.

Bakker, Dorothee, and Andrew Watson. "A Piece in the CO_2 Jigsaw." *Nature* 410 (April 12, 2001): 765–766.

Bala, G., K. Caldeira, A. Mirin, and M. Wickett. "Multicentury Changes to the Global Climate and Carbon Cycle: Results from a Coupled Climate and Carbon Cycle Model." *Journal of Climate* (November 1, 2005): 4531–4544.

Baldwin, Mark P., and Timothy J. Dunkerton. "Stratospheric Harbingers of Anomalous Weather Regimes." *Science* 294 (October 19, 2001): 581–584.

———, David W. J. Thompson, Emily F. Shuckburgh, Warwick A. Norton, and Nathan P. Gillett. "Weather from the Stratosphere?" *Science* 301 (July 18, 2003): 317–318.

Ball, Jeffrey, and Norihiko Shirouzu. "Ford Will Call Earth's Climate a Serious Issue." *Wall Street Journal*, May 2, 2002, A-3, A-19.

Ball, Philip. "Climate Change Set to Poke Holes in Ozone: Arctic Clouds Could Make Ozone Depletion Three Times Worse Than Predicted." *Nature* (March 3, 2004). http://info.nature.com/cgi-bin24/DM/y/eOCBoBfHSKo ChoJVVoAY.

Ballschmiter, Karlheinz. "A Marine Source for Alkyl Nitrates." *Science* 297 (August 16, 2002): 1127–1128.

Banerjee, Neela. "Utility Plans to Put Limits on Its Plants." *New York Times*, May 4, 2001, C-3.

Bange, Hermann. "It's Not a Gas." *Nature* 408 (November 16, 2000): 301–302.

Barbeliuk, Anne. "Warmer Globe Choking Ocean." *Hobart (Australia) Mercury*, March 16, 2002. (Lexis).

Barbraud, C., and H. Weimerskirch. "Emperor Penguins and Climate Change." *Nature* 411 (May 10, 2001): 183–186.

Selected Bibliography

Barford, Carol C., Steven C. Wofsy, Michael L. Goulden, J. William Munger, Elizabeth Hammond Pyle, Shawn P. Urbanski, et al. "Factors Controlling Long- and Short-Term Sequestration of Atmospheric CO_2 in a Mid-Latitude Forest." *Science* 294 (November 23, 2001): 1688–1691.

Barkham, Patrick. "Going Down: Tuvalu, a Nation of Nine Islands—Specks in the South Pacific—Is in Danger of Vanishing, a Victim of Global Warming. As Their Homeland Is Battered by Ferocious Cyclones and Slowly Submerges under the Encroaching Sea, What Will Become of the Islanders?" *London Guardian*, February 16, 2002, 24.

Barnett, Tim P., David W. Pierce, and Reiner Schnur. "Detection of Anthropogenic Climate Change in the World's Oceans." *Science* 292 (April 13, 2001): 270–274.

Baron, Ethan. "Beetles Could Chew Up 80 Per Cent of B.C. Pine: Report: 'Worst-Case Scenario' by 2020 Blamed on Global Warming." *Ottawa Citizen*, September 12, 2004, A-3.

Barrett, Peter. "Cooling of a Continent." *Nature* 421 (January 16, 2003): 221–223.

Barry, Colleen. "Rules for the Kyoto Protocol Adopted in Germany July 23, 2001." Associated Press. www.weather.com/newscenter/topstories/recreation/nparks/010723climateconference.html.

Barry, Leon, George C. Craig, and John Thurburn. "Poleward Heat Transport by the Atmospheric Heat Engine." *Nature* 415 (February 14, 2002): 774–777.

Barta, Patrick, and Rebecca Smith. "Global Surge in Use of Coal Alters Energy Equation." *Wall Street Journal*, November 16, 2004, A-1, A-17.

Bates, Nicholas R., A. Christine Pequignet, Rodney J. Johnson, and Nicolas Gruber. "A Short-Term Sink for Atmospheric CO_2 in Subtropical Mode Water of the North Atlantic Ocean." *Nature* 420 (December 5, 2002): 489–493.

Baumert, Kevin A., ed. *Building on the Kyoto Protocol: Options for Protecting the Climate.* Washington, D.C.: World Resources Institute, 2002.

Baxter, James. "Canada's Forests at Risk of Devastation: Global Warming Could Ravage Tourism, Lumber Industry, Commission Warns." *Ottawa Citizen*, March 5, 2002, A-5.

Beaugrand, Gregory, Keith M. Brander, J. Alistair Lindley, Sami Souissi, and Philip C. Reid. "Plankton Effect on Cod Recruitment in the North Sea." *Nature* 426 (December 11, 2003): 661–664.

———, Philip C. Reid, Frédéric Ibañez, J. Alistair Lindley, and Martin Edwards. "Reorganization of North Atlantic Marine Copepod Biodiversity and Climate." *Science* 296 (May 31, 2002): 1692–1694.

Beck, J. Warren, David A. Richards, R. Lawrence Edwards, Bernard W. Silverman, Peter L. Smart, Douglas J. Donahue, et al. "Extremely Large

Variations of Atmospheric ^{14}C Concentration during the Last Glacial Period." *Science* 292 (June 29, 2001): 2453–2458.

Becker, Jasper. "China's Growing Pains: More Money, More Stuff, More Problems. Any Solutions?" *National Geographic*, March 2004, 68–95.

Becker, L., R. J. Poreda, A. R. Basu, K. O. Pope, T. M. Harrison, C. Nicholson, et al. "Bedout: A Possible End-Permian Impact Crater Offshore of Northwestern Australia." *Science* 304 (June 4, 2004): 1469–1476.

Bell, Jim. "Nunavut Premier Stands Firm on Global Warming." Environment News Service, August 8, 2002. http://ens-news.com/ens/aug2002/2002-08-09-04.asp.

Bell, R. E., M. Studinger, A. A. Tikku, G. K. Clarke, M. M. Gutner, and C. Meertens. "Origin and Fate of Lake Vostok Water Frozen to the Base of the East Antarctic Ice Sheet." *Nature* 416 (March 21, 2002): 307–310.

Bellwood, D. R., T. P. Hughes, C. Folke, and M. Nystrom. "Confronting the Coral Reef Crisis." *Nature* 429 (June 24, 2004): 827–833.

"Ben & Jerry's Flavor Combats Global Warming." Environment News Service, May 2, 2002. http://ens-news.com/ens/may2002/2002L-05-02-09.html#anchor8.

Bengtsson, Lennart. "Hurricane Threats." *Science* 293 (July 20, 2001): 440–441.

———, and Claus U. Hammer. *Geosphere-Biosphere Interactions and Climate*. Cambridge, U.K.: Cambridge University Press, 2001.

Beniston, Martin. *Climatic Change: Implications for the Hydrological Cycle and for Water Management*. Dordrecht, Germany: Kluwer Academic Publishers, 2002.

Bennett, K. D., S. G. Haberle, and S. H. Lumley. "The Last Glacial-Holocene Transition in Southern Chile." *Science* 290 (October 13, 2000): 325–328.

Benson, Simon. "Giant Squid 'Taking Over the World.'" *Sydney Daily Telegraph*, July 31, 2002, 4.

Benton, Michael J. *When Life Nearly Died: The Greatest Mass Extinction of All Time*. London: Thames and Hudson, 2003.

Berger, Andre, and Marie-France Loutre. "Climate: An Exceptionally Long Interglacial Ahead?" *Science* 297 (August 23, 2002): 1287–1288.

Bernard, Harold W., Jr. *Global Warming: Signs to Watch For*. Bloomington: Indiana University Press, 1993.

Betancourt, Julio J. "The Amazon Reveals Its Secrets—Partly." *Science* 290 (December 22, 2000): 2274–2275.

Betsill, Michelle M. "Impacts of Stratospheric Ozone Depletion." In *Handbook of Weather, Climate, and Water: Atmospheric Chemistry, Hydrology, and Societal Impacts*, ed. Thomas D. Potter and Bradley R. Colman, 913–923. Hoboken, NJ: Wiley Interscience, 2003.

Betts, Richard A. "Offset of the Potential Carbon Sink from Boreal Forestation by Decreases in Surface Albedo." *Nature* 408 (November 9, 2000): 187–190.

Selected Bibliography

Biggin, Susan. "Venice Preservation: Climate Change Data Prompt New Review." *Science* 292 (April 6, 2001): 28.

"Billions of People May Suffer Severe Water Shortages as Glaciers Melt: World Wildlife Fund."Agence France Presse, November 27, 2003. (Lexis).

Bindschadler, Robert A., Matt A. King, Richard B. Alley, Sridhar Anandakrishnan, and Laurence Padman. "Tidally Controlled Stick-Slip Discharge of a West Antarctic Ice." *Science* 301 (August 22, 2003): 1087–1089.

"Biomass Burning Boosts Stratospheric Moisture." Environment News Service, February 20, 2002. http://ens-news.com/ens/feb2002/2002L-02-20-09.html.

Bishop, James K. B., Russ E. Davis, and Jeffrey T. Sherman. "Robotic Observations of Dust Storm Enhancement of Carbon Biomass in the North Pacific." *Science* 298 (October 25, 2002): 817–821.

"Bitter Pill: The Northward Spread of the Okinawan Goya, Warm-Weather." Asahi News Service (Japan), January 29, 2003. (Lexis).

Black, David E. "The Rains May Be A-Comin'." *Science* 297 (July 26, 2002): 528–529.

Bleach, Stephen. "The Naked Truth." *London Sunday Times*, August 24, 2003, Travel Section, 1.

Bloomfield, Janine, and Sherry Showell. *Global Warming: Our Nation's Capital at Risk*. Environmental Defense Fund, 1997. www.edf.org/pubs/Reports/WashingtonGW/index.html.

Blunier, Thomas, and Edward J. Brook. "Timing of Millennial-Scale Climate Change in Antarctica and Greenland during the Last Glacial Period." *Science* 291 (January 5, 2001): 109–112.

Bobylev, Leonid P., Kiril Ya. Kondratyev, and Ola M. Johannessen. *Arctic Environment Variability in the Context of Global Change*. New York: Springer, 2004.

Bodhaine, Barry, Ellsworth Dutton, and Renee Tatusko. "Assessment of Ultraviolet (UV) Variability in the Alaskan Arctic." Cooperative Institute for Arctic Research, University of Alaska and NOAA, March 6, 2001. www.cifar.uaf.edu/ario0/bodhaine.html.

Bolin, Bert, John T. Houghton, Gylvan Meira Filho, Robert T. Watson, M. C. Zinyowera, James Bruce, et al. *Intergovernmental Panel on Climate Change. Second Assessment Synthesis of Scientific-Technical Information Relevant to Interpreting Article 2 of the United Nations Framework Convention on Climate Change*. Rome, December 11–15, 1995. www.unep.ch/ipcc/pub/sarsyn.htm.

———, and Haroon S. Kheshgi. "On Strategies for Reducing Greenhouse Gas Emissions." *Proceedings of the National Academy of Sciences* 98 (9) (April 24, 2001): 4850–4854.

Bond, Gerard, Bernd Kromer, Juerg Beer, Raimund Muscheler, Michael N. Evans, William Showers, et al. "Persistent Solar Influence on North Atlantic Climate during the Holocene." *Science* 294 (December 7, 2001): 2130–2136.

———, W. Showers, M. Cheseby, R. Lotti, P. Almasi, P. deMenocal, et al. "A Pervasive, Millennial-Scale in North Atlantic Holocene and Glacial Climates." *Science* 278 (1997): 1257–1266.

Borenstein, Seth. "Odds on Global Warming Issued." *Montreal Gazette*, July 20, 2001, A-1.

———. "Scientists Worry about Evidence of Melting Arctic Ice." Knight-Ridder News Service, *Seattle Times*, February 18, 2005, A-6.

Bosch, Bosch. "Great Balls of Ice!" *Science* 297 (August 2, 2002): 765.

Boswell, Randy. "Gigantic Ice Shelf Breaks Up." CanWest News Service, *Calgary Herald*, September 23, 2003, A-1.

———. "Southern Butterfly's Trek North Cited as Proof of Global Warming." *Edmonton Journal*, November 19, 2004, A-7.

Both, C., and M. E. Visser. "Adjustment to Climate Change Is Constrained by Arrival Date in a Long-Distance Migrant Bird." *Nature* 411 (May 17, 2001): 296–298.

Bourne, Joel K., Jr. "The Big Uneasy." *National Geographic*, October 2004, 88–105.

Bousquet, Philippe, Philippe Peylin, Philippe Ciais, Corinne Le Quéré, Pierre Friedlingstein, and Pieter P. Tans. "Regional Changes in Carbon Dioxide Fluxes of Land and Oceans since 1980." *Science* 290 (November 17, 2000): 1342–1346.

Bowen, Gabriel J., David J. Beerling, Paul L. Koch, James C. Zachos, and Thomas Quattlebaum. "A Humid Climate State during the Palaeocene/Eocene Thermal Maximum." *Nature* 432 (November 25, 2004): 495–499.

Bowen, Jerry. "Dramatic Climate Change in Alaska." CBS News Transcripts, CBS Morning News, August 29, 2002. (Lexis).

Bowen, Mark. *Thin Ice: Unlocking the Secrets of Climate in the World's Highest Mountains.* New York: Henry Holt, 2005.

Bower, A. S., B. Le Cann, T. Rossby, W. Zenk, J. Gould, K. Speer, et al. "Directly Measured Mid-Depth Circulation in the Northeastern North Atlantic Ocean." *Nature* 419 (October 10, 2002): 603–607.

Bowman, Lee. "Soot Could Be Causing a Lot of Bad Weather." Scripps-Howard News Service, September 26, 2002. (Lexis).

Boxall, Bettina. "Epic Droughts Possible, Study Says: Tree Ring Records Suggest That If Past Is Prologue, Global Warming Could Trigger Much Longer Dry Spells Than the One Now in West, Scientists Say." *Los Angeles Times*, October 8, 2004, A-17.

Boyd, Philip W. "Ironing Out Algal Issues in the Southern Ocean." *Science* 304 (April 16, 2004): 396–397.

———, Andrew J. Watson, Cliff S. Law, Edward R. Abraham, Thomas Trull, Rob Murdoch, et al. "A Mesoscale Phytonplankton Bloom in the Polar Southern Ocean Stimulated by Iron Fertilization." *Nature* 407 (October 12, 2000): 695–702.

Boyd, Robert S. "Earth Warming Could Open Up a Northwest Passage." Knight-Ridder Newspapers, *Pittsburgh Post-Gazette*, November 11, 2002, A-1.

———. "Rising Tides Raises Questions: Satellites Will Provide Exact Measurements." Knight-Ridder Newspapers, *Pittsburgh Post-Gazette*, December 9, 2001, A-3.

Boyle, Ed. "Oceanography: Oceanic Salt Switch." *Science* 298 (November 29, 2002): 1724–1725.

Bradley, Raymond S., Malcolm K. Hughes, and Henry F. Diaz. "Climate in Medieval Time." *Science* 302 (October 17, 2003): 404–405.

Bradshaw, William E., and Christina M. Holzapfel. "Genetic Shift in Photoperiodic Response Correlated with Global Warming." *Proceedings of the National Academy of Sciences* 98 (25) (December 4, 2001): 14509–14515.

Bradsher, Keith. "China Prospering but Polluting: Dirty Fuels Power Economic Growth." *New York Times* in *International Herald-Tribune*, October 22, 2003, 1.

———. "Ford Tries to Burnish Image by Looking to Cut Emissions." *New York Times*, May 4, 2001, C-3.

Brasseur, Guy P., Anne K. Smith, Rashid Khosravi, Theresa Huang, Stacy Walters, Simon Chabrillat, et al. "Natural and Human-Induced Pertubations in the Middle Atmosphere: A Short Tutorial." In *Atmospheric Science across the Stratopause*, ed. David E. Siskind et al. Washington, D.C.: American Geophysical Union, 2000.

"Breakaway Bergs Disrupt Antarctic Ecosystem." Environment News Service, May 9, 2002. http://ens-news.com/ens/may2002/2002L-05-09-01.html.

Breed, Allen G. "New Orleans Evacuation Picking Up Steam, but Help Comes Too Late for Untold Number." Associated Press, September 3, 2005. (Lexis).

Brèon, Francois-Marie, Didier Tanrè, and Sylvia Generoso. "Aerosol Effect on Cloud Droplet Size Monitored from Satellite." *Science* (February 1, 2002): 834–838.

British Broadcasting Corporation. "Severe Loss to Arctic Ozone." British Broadcasting Corporation News, April 5, 2000. http://news.bbc.co.uk/hi/english/sci/tech/newsid_702000/702388.stm.

Broecker, Wallace S. "Are We Headed for a Thermohaline Catastrophe?" In *Geological Perspectives of Global Climate Change*, Studies in Geology #17, ed.

Lee C. Gerhard, William E. Harrison, and Bernold M. Hanson, 83–95. Tulsa, OK: American Association of Petroleum Geologists, 2001.

———. "Does the Trigger for Abrupt Climate Change Reside in the Ocean or in the Atmosphere?" *Science* 300 (June 6, 2003): 1519–1522.

———. "Fate of Fossil Fuel Carbon Dioxide and the Global Carbon Budget." *Science* 206 (1979): 409–418.

———. "Future Global Warming Scenarios." Letter to the Editor. *Science* 304 (April 16, 2004): 388.

———. "Thermohaline Circulation: The Achilles Heel of Our Climate System: Will Man-Made CO_2 Upset the Current Balance?" *Science* 278 (1997): 1582–1588.

———. "Unpleasant Surprises in the Greenhouse?" *Nature* 328 (1987): 123–126.

———. "Was the Medieval Warm Period Global?" *Science* 291 (February 23, 2001): 1497–1499.

———, and Sidney Hemming. "Paleoclimate: Climate Swings Come into Focus." *Science* 294 (December 14, 2001): 2308–2309.

Brook, Edward J. "Tiny Bubbles Tell All." *Science* 310 (November 25, 2005): 1285–1287.

Brooke, James. "'Heat Island' Tokyo Is Global Warming's Vanguard." *New York Times*, August 13, 2002, A-3.

Brown, Amanda. "New Figures Show Fall in Greenhouse Gas Emissions." Press Association (United Kingdom), March 27, 2003. (Lexis).

Brown, DeNeen L. "Greenland's Glaciers Crumble: Global Warming Melts Polar Ice Cap into Deadly Icebergs." *Washington Post*, October 13, 2002, A-30.

———. "Hamlet in Canada's North Slowly Erodes: Arctic Community Blames Global Warming as Permafrost Starts to Melt and Shoreline." *Washington Post*, September 13, 2003, A-14.

———. "Waking the Dead, Rousing Taboo: In Northwest Canada, Thawing Permafrost Is Unearthing Ancestral Graves." *Washington Post*, October 17, 2001, A-27.

Brown, Donald A. *American Heat: Ethical Problems with the United States' Response to Global Warming.* Lanham, MD: Rowman & Littlefield, 2002.

Brown, Kathryn. "Pacific Salmon Run Hot and Cold." *Science* 290 (October 27, 2000): 685–686.

Brown, Paul. "Analysis: Blair Sets Out Far-Reaching Vision but Where Are the Practical Policies?" *London Guardian*, February 25, 2003, 13.

———. "Climate Expert Accuses Prime Minister of Cowardice." *London Guardian*, July 28, 2003, 2.

———. "Geographers' Conference: Ice Field Loss Puts Alpine Rivers at Risk: Global Warming Warning to Europe." *London Guardian*, January 5, 2002, 9.

———. "Global Warming Is Killing Us Too, Say Inuit." *London Guardian*, December 11, 2003, 14.

———. "Global Warming: It's with Us Now: Six Dead as Storms Bring Chaos throughout the Country." *London Guardian*, October 31, 2000, 1.

———. "Global Warming Kills 150,000 a Year: Disease and Malnutrition the Biggest Threats, United Nations Organisations Warn at Talks on Kyoto." *London Guardian*, December 12, 2003, 19.

———. "Global Warming: Worse Than We Thought." *World Press Review*, February 1999, 44.

———. "Islands in Peril Plead for Deal." *London Guardian*, November 24, 2000, 21.

———. "Melting Permafrost Threatens Alps: Communities Face Devastating Landslides from Unstable Mountain Ranges." *London Guardian*, January 4, 2001, 3.

———. [No headline]. *London Guardian*, November 1, 2000, 1.

———. "Scientists Warn of Himalayan Floods: Global Warming Melts Glaciers and Produces Many Unstable Lakes." *London Guardian*, April 17, 2002, 13.

———, and Tony Sutton. "Global Warming Brings New Cash Crop to West Country as Rising Water Temperatures Allow Valuable Shellfish to Thrive." *London Guardian*, December 10, 2002, 8.

Browne, Anthony. "Canute Was Right! Time to Give Up the Coast." *London Times,* October 11, 2002, 8.

———. "How Climate Change Is Killing Off Rare Animals: Conservationists Warn That Nature's 'Crown Jewels' Are Facing Ruin." *London Observer*, February 10, 2002, 15.

———, and Paul Simons. "Euro-Spiders Invade as Temperature Creeps Up." *London Times*, December 24, 2002, 8.

"Brush Fires Collapsing Bear Dens." Canadian Press in *Calgary Sun*, November 2, 2002, 18.

Bryden, Harry L., Hannah R. Longworth, and Stuart A. Cunningham. "Slowing of the Atlantic Meridional Overturning Circulation at 25° North." *Nature* 438 (December 1, 2005): 655–657.

Buddemeier, Robert W., John R. Ware, Toby A. Gardner, Isabelle M. Cùtè, Jennifer A. Gill, Alastair Grant, et al. "Coral Reef Decline in the Caribbean." Letter to the Editor. *Science* 302 (October 17, 2003): 391–392.

Bueckert, Dennis. "Climate Change Could Bring Malaria, Dengue Fever to Southern Ontario, Says Report." Canadian Press in *Ottawa Citizen*, October

23, 2002. www.canada.com/news/story.asp?id={B019135A-4FD8-4536-A2F0-908F13560CB2.

———. "Climate Change Linked to Ill Health in Children." Canadian Press in *Montreal Gazette*, June 2, 2001, A-2.

———. "Forest Fires Taking Toll on Climate: CO_2 from Increased Burning Could Overtake Fossil Fuels as a Source of Global Warming, Prof Warns." Canadian Press in *Edmonton Journal*, September 19, 2002, A-3.

Buencamino, Manuel. "Coming Catastrophe?" *BusinessWorld*, June 14, 2004, 21.

Buesseler, Ken O., John E. Andrews, Steven M. Pike, and Matthew A. Charette. "The Effects of Iron Fertilization on Carbon Sequestration in the Southern Ocean." *Science* 304 (April 16, 2004): 414–417.

———, and Philip W. Boyd. "Will Ocean Fertilization Work?" *Science* 300 (April 4, 2003): 67–68.

Buist, Steve. "What's Still Green but Should Be Red All Over?" *Hamilton (Ontario) Spectator*, October 8, 2002, A-8.

Bunting, Madeleine. "Confronting the Perils of Global Warming in a Vanishing Landscape: As Vital Talks Begin at The Hague, Millions Are Already Suffering the Consequences of Climate Change." *London Guardian*, November 14, 2000, 1.

Burnett, Adam W., Matthew E. Kirby, Henry T. Mullins, and William P. Patterson. "Increasing Great Lake-Effect Snowfall during the Twentieth Century: A Regional Response to Global Warming?" *Journal of Climate* 16 (21) (November 1, 2003): 3535–3542.

Burnham, Michael. "Scientists Link Global Warming with Increasing Marine Diseases." Greenwire, October 7, 2003. (Lexis).

Burns, Stephen J., Dominik Fleitmann, Albert Matter, Jan Kramers, and Abdulkarim A. Al-Subbary. "Indian Ocean Climate and an Absolute Chronology over Dansgaard/Oeschger Events 9 to 13." *Science* 301 (September 5, 2003): 1365–1367.

Burroughs, William. *Climate into the 21st Century*. New York: Cambridge University Press, 2003.

Burt, Christopher C. *Extreme Weather: A Guide and Record Book*. New York: Norton, 2004.

Bush, Mark B., Miles R. Silman, and Dunia H. Urrego. "48,000 Years of Climate and Forest Change in a Biodiversity Hot Spot." *Science* 303 (February 6, 2004): 827–829.

"Bush Says Administration Has Not Changed Stance on Global Warming." *The Frontrunner*, August 27, 2004. (Lexis).

Butchart, Neal, and Adam A. Scaife. "Removal of Chlorofluorocarbons by Increased Mass Exchange between the Stratosphere and the Troposphere in a Changing Climate." *Nature* 410 (April 12, 2001): 799–802.

Byers, Stephen, and Olympia Snowe, co-chairs, International Climate Change Task Force. *Meeting the Climate Challenge: Recommendations of the International Climate Change Task Force.* London: Institute for Public Policy Research, January 2005.

Cabanes, Cecile, Anny Cazenave, and Christian Le Provost. "Sea Level Rise during Past 40 Years Determined from Satellite and In Situ Observations." *Science* 294 (October 26, 2001): 840–842.

Caillon, Nicolas, Jeffrey P. Severinghaus, Jean Jouzel, Jean-Marc Barnola, Jiancheng Kang, and Volodya Y. Lipenkov. "Timing of Atmospheric CO_2 and Antarctic Temperature Changes across Termination III." *Science* 299 (March 14, 2003): 1728–1731.

Calamai, Peter. "Alert over Shrinking Ozone Layer." *Toronto Star*, March 18, 2002, A-8.

———. "Atlantic Water Changing: Scientists." *Toronto Star*, June 21, 2001, A-18.

———. "Global Warming Threatens Reindeer." *Toronto Star*, December 23, 2002, A-23.

Caldeira, Ken, Atul K. Jain, and Martin I. Hoffert. "Climate Sensitivity Uncertainty and the Need for Energy without CO_2 Emissions." *Science* 299 (March 28, 2003): 2052–2054.

———, and Michael E. Wickett. "Oceanography: Anthropogenic Carbon and Ocean pH." *Nature* 425 (September 25, 2003): 365.

Callendar, G. D. "The Artificial Production of Carbon Dioxide and Its Influence on Temperature." *Quarterly Journal of the Royal Meteorological Society* 64 (1938): 223–240.

Calvin, William H. *A Brain for All Seasons: Human Evolution and Abrupt Climate Change.* Chicago: University of Chicago Press, 2002.

Campbell, Duncan. "Greenhouse Melts Alaska's Tribal Ways: As Climate Talks Get Under Way in Bonn Today, Some Americans Are Ruing the Warming Their President Chooses to Ignore." *London Guardian*, July 16, 2001, 11.

Cane, Mark A., and Michael Evans. "Do the Tropics Rule?" *Science* 290 (November 10, 2000): 1107–1108.

Cane, M. A., and P. Molnar. "Closing of the Indonesian Seaway as a Precursor to East Aridification around 3 to 4 Million Years Ago." *Nature* 411 (May 10, 2001): 157–160.

Capella, Peter. "Disasters Will Outstrip Aid Effort as World Heats Up: Rich States Could Be Sued as Voluntary Assistance Falters, Red Cross Says." *London Guardian*, June 29, 2001, 15.

———. "Europe's Alps Crumbling: Glaciers Melting in Heatwave." Agence France Presse, August 7, 2003. (Lexis).

Capiello, Dina. "Adirondacks Climate Growing Hotter Faster." *Albany Times-Union*, September 21, 2002.

"Carbon Pollution Wreaking Havoc with Amazonian Forest." Agence France Presse, March 10, 2004. (Lexis).

"Carbon Sinks Cannot Keep Up with Emissions." Environment News Service, May 16, 2002. http://ens-news.com/ens/may2002/2002L-05-16-09.html# anchor3.

Carey, John, and Sarah R. Shapiro. "Consensus Is Growing among Scientists, Governments, and Business That They Must Act Fast to Combat Climate Change. This Has Already Sparked Efforts to Limit CO_2 Emissions. Many Companies Are Now Preparing for a Carbon-Constrained World." *Business Week*, August 16, 2004. (Lexis).

Carpenter, Betsy. "Feeling the Sting: Warming Oceans, Depleted Fish Stocks, Dirty Water—They Set the Stage for a Jellyfish Invasion." *U.S. News and World Report*, August 16, 2004, 68–69.

Carslaw, K. S., R. G. Harrison, and J. Kirkby. "Cosmic Rays, Clouds, and Climate." *Science* 298 (November 29, 2002): 1732–1737.

Caspersen, John P., Stephen W. Pacala, Jennifer C. Jenkins, George C. Hurtt, Paul R. Moorcroft, and Richard A. Birdsey. "Contributions of Land-Use History to Carbon Accumulation in U.S. Forests." *Science* 290 (November 10, 2000): 1148–1151.

Catholic Bishops, U.S. Conference. "Global Climate Change: A Plea for Dialogue, Prudence, and the Common Good: A Statement of the U.S. Catholic Bishops." Ed. William P. Fay, June 15, 2001. www.ncrlc.com/climideas.html.

"Century of Human Impact Warms Earth's Surface." Environment News Service, January 24, 2002. http://ens-news.com/ens/jan2002/2002L-01-24-09.html.

Chambers, Kevin. "Fewer Frosty Days." Weather.com, January 17, 2001. www .weather.com/weather_center/full_story/full3.html.

Chameides, William L., and Michael Bergin. "Climate Change: Soot Takes Center Stage." *Science* 297 (September 27, 2002): 2214–2215.

Chang, Ching-Cheng, Robert Mendelsohn, and Daigee Shaw, eds. *Global Warming and the Asian Pacific*. Cheltenham, U.K.: Edward Elgar, 2003.

Chang, Kenneth. "Arctic Ice Is Melting at Record Level, Scientists Say." *New York Times*, December 8, 2002, A-40.

———. "The Melting (or Freezing) of Antarctica: Deciphering Contradictory Climate Patterns Is Largely a Matter of Ice." *New York Times*, April 2, 2002, F-1.

———. "Ozone Hole Is Now Seen as a Cause for Antarctic Cooling." *New York Times*, May 3, 2002, A-16.

"Changes in Climate Bring Hops Northward." *Glasgow (Scotland) Herald*, September 29, 2000, 13.

Chapin, F. S., III, M. Sturm, M. C. Serreze, J. P. McFadden, J. R. Key, A. H. Lloyd, A. D. McGuire, T. S. Rupp, A. H. Lynch, J. P. Schimel, J. Beringer, W. L. Chapman, H. E. Epstein, E. S. Euskirchen, L. D. Hinzman, G. Jia, C.-L. Ping, K. D. Tape, C.D.C. Thompson, D. A. Walker, and J. M. Welker. "Role of Land-Surface Changes in Arctic Summer Warming." *Science* 310 (October 28, 2005): 657–660.

Chapman, James. "Early Spring Misery for 12 Million Hay Fever Sufferers." *London Daily Mail*, February 4, 2003, 23.

Chapman, W. L., and J. E. Walsh. "Recent Variations of Sea Ice and Air Temperature in High Latitudes." *Bulletin of the Meteorological Society of America* 73 (1993): 34–47.

Charlson, Robert J. "Giants' Footprints in the Greenhouse: The Seeds of Our Understanding of Global Warming Were Sewn by Early Heroes." *Nature* 401 (October 21, 1999): 741–742.

———, John H. Seinfeld, Athanasios Nenes, Markku Kulmala, Ari Laaksonen, and M. Cristina Facchini. "Reshaping the Theory of Cloud Formation." *Science* 292 (June 15, 2001): 2025–2026.

Chase, Marilyn. "As Virus Spreads, Views of West Nile Grow Even Darker." *Wall Street Journal*, October 14, 2004, A-1, A-10.

Chase, Steven. "Our Water Is at Risk, Climate Study Finds." *Toronto Globe and Mail*, August 13, 2002, A-1. www.globeandmail.com/servlet/ArticleNews/ PEstory/TGAM/20020813/UENVIN/national/national/national_temp/6/ 6/23/.

Chen, Junye, Barbara E. Carlson, and Anthony D. Del Genio. "Evidence for Strengthening of the Tropical General Circulation in the 1990s." *Science* 295 (February 1, 2002): 838–841.

Chengappa, Raj. "The Monsoon: What's Wrong with the Weather?" *India Today*, August 12, 2002, 40.

Chin, Gilbert. "Editors' Choice: No Deepwater Slowdown?" *Science* 293 (July 27, 2001): 575.

"China to Experience Continuous 'Greenhouse Effect' in Next 50 to 100 Years." Xinhua News Service, September 7, 2002. (Lexis).

Chisholm, Sallie W. "Stirring Times in the Southern Ocean." *Nature* 407 (October 12, 2000): 685–686.

Choi, Charles. "Rainforests Might Speed Up Global Warming." United Press International, April 21, 2003. (Lexis).

Christensen, Jens H., and Ole B. Christensen. "Severe Summertime Flooding in Europe." *Nature* 421 (February 20, 2003): 805.

Christensen, Torben R., Torbjörn Johansson, H. Jonas Kerman, Mihail Mastepanov, Nils Malmer, Thomas Friborg, et al. "Thawing Sub-Arctic

Permafrost: Effects on Vegetation and Methane Emissions." *Geophysical Research Letters* 31 (4) (February 20, 2004). www.agu.org/pubs/current/gl .shtml/index.php?month=February.

Christianson, Gale E. *Greenhouse: The 200-Year Story of Global Warming.* New York: Walker and Company, 1999.

Christie, Maureen. *The Ozone Layer: A Philosophy of Science Perspective.* Cambridge, U.K.: Cambridge University Press, 2001.

Christy, John R., Roy W. Spencer, B. D. Santer, T. M. L. Wigley, G. A. Meehl, M. F. Wehner, et al. "Reliability of Satellite Data Sets." Letter to the Editor. *Science* 301 (August 22, 2003): 1045.

Church, John A. "Climate Change: How Fast Are Sea Levels Rising?" *Science* 294 (October 26, 2001): 802–803.

Chylek, Petr, Jason E. Box, and Glen Lesins. "Global Warming and the Greenland Ice Sheet." *Climatic Change* 63 (2004): 201–221.

Ciborowski, Peter. "Sources, Sinks, Trends, and Opportunities." In *The Challenge of Global Warming*, ed. Edwin Abrahamson, 213–230. Washington, D.C.: Island Press, 1989.

Cifuentes, Luis, Victor H. Borja-Aburto, Nelson Gouveia, George Thurston, and Devra Lee Davis. "Hidden Health Benefits of Greenhouse Gas Mitigation." *Science* 252 (August 17, 2001): 1257–1259.

Clark, D. A., S. C. Piper, C. D. Keeling, and D. B. Clark. "Tropical Rain Forest Tree Growth and Atmospheric Carbon Dynamics Linked to Interannual Temperature Variation during 1984–2000." *Proceedings of the National Academy of Sciences* 100 (10) (May 13, 2003): 5852–5857.

Clark, Peter U., Shawn J. Marshall, Garry K. C. Clarke, Steven W. Hostetler, Joseph M. Licciardi, and James T. Teller. "Freshwater Forcing of Abrupt Climate Change during the Last Glaciation." *Science* 293 (July 13, 2001): 283–287.

———, A. Marshall McCabe, Alan C. Mix, and Andrew J. Weaver. "Rapid Rise of Sea Level 19,000 Years Ago and Its Global Implications." *Science* 304 (May 21, 2004): 1141–1144.

———, J. X. Mitrovica, G. A. Milne, and M. E. Tamisiea. "Sea-Level Fingerprinting as a Direct Test for the Source of Global Meltwater Pulse." *Science* 295 (March 29, 2002): 2438–2441.

———, N. G. Pisias, T. F. Stocker, and A. J. Weaver. "The Role of the Thermohaline Circulation in Abrupt Climate Change." *Nature* 415 (February 21, 2002): 863–868.

———, G. O. Seltzer, D. T. Rodbell, P. A. Baker, S. C. Fritz, P. M. Tapia, et al. "Early Deglaciation in the Tropical Andes." *Science* 298 (October 4, 2002): 7.

Clark, Ross. "Rain, Rain Come Again: The Long Dry Spell Is Making Gardeners Anxious." *London Daily Telegraph*, May 17, 2003, 15.

Clarke, Garry, David Leverington, James Teller, and Arthur Dyke. "Enhanced: Superlakes, Megafloods, and Abrupt Climate Change." *Science* 301 (August 15, 2003): 922–923.

Clarke, Tom. "Boiling Seas Linked to Mass Extinction: Methane Belches May Have Catastrophic Consequences." *Nature*, August 22, 2003. http://info.nature.com/cgi-bin24/DM/y/eLodoBfHSKoChoDYyoAL.

Claussen, Eileen, ed. *Climate Change: Science, Strategies, and Solutions.* Arlington, VA: Pew Center on Global Climate Change, 2001.

———. "An Effective Approach to Climate Change." *Science* 306 (October 29, 2004): 816.

Clavel, Guy. "Global Warming Makes Polar Bears Sweat." Agence France Presse, November 3, 2002.

Clayton, Chris. "U.S.D.A. Will Offer Incentives for Conserving Carbon in Soil." *Omaha World-Herald*, June 7, 2003, D-1, D-2.

Clift, Peter, and Karen Bice. "Earth Science: Baked Alaska." *Nature* 419 (September 12, 2002): 129–130.

"Climate Change Acceleration Will Push Claims Bills Higher." *Insurance Day*, March 4, 2004, 1.

"Climate Change and the Financial Services Industry." United Nations Environment Programme Finance Initiative, No date. www.unepfi.net.

"Climate, Changing Agricultural Methods Affect Lake Erie's Health." Environment News Service, April 16, 2002. http://ens-news.com/ens/apr2002/2002L-04-16-09.html.

"Climate-Related Perils Could Bankrupt Insurers." Environment News Service, October 7, 2002. http://ens-news.com/ens/oct2002/2002-10-07-02.asp.

"Climate Warms Twice as Fast." British Broadcasting Corporation Monitoring Asia-Pacific, January 6, 2003. (Lexis).

"Climatology: Warmer Weathering." *Science* 292 (May 4, 2001): 811.

Cline, William R. *The Economics of Global Warming.* Washington, D.C.: Institute for International Economics, 1992.

"Clouds, but No Silver Lining." *London Guardian*, January 24, 2002.

Clover, Charles. "Climate Conference Pounds 1,000 Grant to Switch to 'Green' Cars." *London Daily Telegraph*, November 21, 2000, 9.

———. "Geographers' Conference: Alps May Crumble as Permafrost Melts." *London Telegraph*, January 4, 2001, 12.

———. "Global Warming 'Is Driving Fish North.'" *London Daily Telegraph*, May 31, 2002, 14.

———. "2002 'Warmest for 1,000 Years.'" *London Daily Telegraph*, April 26, 2002, 1.

———, and David Millward. "Future of Cheap Flights in Doubt: Ban New Runways and Raise Fares, Say Pollution Experts." *London Daily Telegraph*, November 30, 2002, 1, 4.

Coale, Kenneth H., Kenneth S. Johnson, Francisco P. Chavez, Ken O. Buesseler, Richard T. Barber, Mark A. Brzezinski, et al. "Southern Ocean Iron Enrichment Experiment: Carbon Cycling in High- and Low-Si Waters." *Science* 304 (April 16, 2004): 408–414.

"Coastal Gulf States Are Sinking." Environment News Service, April 21, 2003. http://ens-news.com/ens/apr2003/2003-04-21-09.asp#anchor4.

Cole, Julia. "A Slow Dance for El Niño." *Science* 291 (February 23, 2001): 1496–1497.

Collier, Michael, and Robert H. Webb. *Floods, Drought, and Climate Change.* Tucson: University of Arizona Press, 2002.

Collins, Simon. "Birds Starve in Warmer Seas." *New Zealand Herald*, November 14, 2002. (Lexis).

Collins, Sinead, and Graham Bell. "Phenotypic Consequences of 1,000 Generations of Selection at Elevated CO_2 in a Green Alga." *Nature* 431 (September 30, 2004): 566–569.

Comiso, Josefino C. "A Rapidly Declining Perennial Sea Ice Cover in the Arctic." *Geophysical Research Letters* 29 (20) (October 18, 2002): 1956–1960.

———. "Warming Trends in the Arctic from Clear Sky Satellite Observations." *Journal of Climate* 16 (21) (November 1, 2003): 3498–3510.

"Conferences Tackle Key Issues in Air Conditioning, Refrigeration." AScribe Newswire, June 23, 2004. (Lexis).

Connor, Steve. "Britain Could Become as Cold as Moscow." *London Independent*, June 21, 2001, 14.

———. "Catastrophic Climate Change 90 Per Cent Certain." *London Independent*, July 20, 2001, 15.

———. "El Niño's Rise May Be Linked to Pacific Current Slowdown." *London Independent*, February 7, 2002.

———. "Global Warming Is Choking the Life Out of Lake Tanganyika." *London Independent*, August 14, 2003. (Lexis).

———. "Global Warming Is Twice as Bad as Previously Thought." *London Independent*, January 27, 2005, 10.

———. "Global Warming May Wipe Out a Fifth of Wild Flower Species, Study Warns." *London Independent*, June 17, 2003. (Lexis).

———. "Malaria Could Become Endemic Disease in U.K." *London Independent*, September 12, 2001, 14.

———. "Meltdown: Arctic Wildlife Is on the Brink of Catastrophe: Polar Bears Could Be Decades from Extinction." *London Independent*, November 11, 2004. (Lexis).

———. "Peat Bog Gases Accelerate Global Warming." *London Independent*, July 8, 2004, 9.

———. "Polar Sea Ice Could Be Gone by the End of the Century." *London Independent*, March 10, 2003, 5.

———. "Strangers in the Seas: Exotic Marine Species Are Turning Up Unexpectedly in the Cold Waters of the North Atlantic." *London Independent*, August 5, 2002, 12–13.

———. "World's Wildlife Shows Effects of Global Warming." *London Independent*, March 28, 2002, 11.

"Conservation of Arctic Flora and Fauna: Arctic Climate Impact Assessment. An Assessment of Consequences of Climate Variability and Change and the Effects of Increased UV in the Arctic Region: A Draft Implementation Plan." United Nations Environment Programme, October 22, 1999. Constantineau, Bruce. "Weather Wreaking Havoc on B.C. Farms: Global Warming Has Increased the Intensity and Frequency of Weather Events, Making It Difficult for Farmers to Compete in the Marketplace." *Vancouver Sun*, March 3, 2003, D-3.

"Contrails Linked to Temperature Changes." Environment News Service, August 8, 2002. http://ens-news.com/ens/aug2002/2002-08-08-09.asp#anchor4.

Cook, Edward R., Connie A. Woodhouse, C. Mark Eakin, David M. Meko, and David W. Stahle. "Long-Term Aridity Changes in the Western United States." *Science* 306 (November 5, 2004): 1015–1018.

Cooke, Robert. "Global Warming Is in the Air." *Newsday*, September 20, 2002, A-34.

———. "Is Global Warming Making Earth Greener?" *Newsday*, September 11, 2001, C-3.

———. "Scientists: Pacific Slower at Surface: Data May Help Explain Trend in El Niño Events." *Newsday*, February 7, 2002, A-46.

———. "Waters Reflect Weather Trend: Study Finds Warming Effects." *Newsday*, December 18, 2003, A-2.

Cookson, Clive. "Explorer Gives Up Trek to North Pole as Global Warming Melts Sea Ice." *London Financial Times*, May 18, 2002, 5.

———. "Global Warming Triggers Epidemics in Wildlife." *London Financial Times*, June 21, 2002, 4.

———, and Victoria Griffith. "Blame for Flooding May Be Misplaced: Climate Change Global Warming May Not Be the Reason for Recent Heavy Rainfall in Europe and Asia." *London Financial Times*, August 15, 2002, 6.

Cookson, Craig, and David Firn. "Breeding Bugs That May Help Save the World: Craig Venter Has Found a Large Project to Follow the Human Genome." *London Financial Times*, September 28, 2002, 11.

Corkish, R. "A Power That's Clean and Bright." Review of *Clean Energy from Photovoltaics*, ed. Mary D. Archer and Robert Hill. *Nature* 416 (April 18, 2002): 680–681.

Cowen, Robert C. "In Polar Waters, a Surge in Temperatures Takes Scientists by Surprise." *Christian Science Monitor*, 14.

———. "Into the Cold? Slowing Ocean Circulation Could Presage Dramatic—and Chilly—Climate Change." *Christian Science Monitor*, September 26, 2002, 14.

———. "One Large, Overlooked Factor in Global Warming: Tropical Forest Fires." *Christian Science Monitor*, November 7, 2002, 14.

Cox, Peter M., Richard A. Betts, Chris D. Jones, Steven A. Spall, and Ian J. Totterdell. "Acceleration of Global Warming Due to Carbon-Cycle Feedbacks in a Coupled Climate Model." *Nature* 408 (November 9, 2000): 184–187.

Coy, Peter. "The Hydrogen Balm? Author Jeremy Rifkin Sees a Better, Post-Petroleum World." *Business Week*, September 30, 2002, 83.

Cramb, Auslan. "Highland River Salmon 'On Verge of Extinction.'" *London Daily Telegraph*, July 15, 2002, 7.

Crenson, Matt. "Louisiana Sinking: One State's Environmental Nightmare Could Become Common Problem." Associated Press, August 10, 2002. (Lexis).

Crilly, Rob. "2050 to Be Good Year for Scottish Wine: Global Warming Will Bring Grapes North." *Glasgow Herald*, November 20, 2002, 11.

Crowley, Thomas J. "Paleoclimate: Cycles, Cycles Everywhere." *Science* 295 (February 22, 2002): 1473–1474.

———, and Robert A. Berner. "Enhanced: CO_2 and Climate Change." *Science* 292 (May 4, 2001): 870–872.

Croxall, J. P., P. N. Trathan, and E. J. Murphy. "Environmental Change and Antarctic Seabird Populations." *Science* 297 (August 30, 2002): 1510–1514.

Crutzen, Paul J. "The Antarctic Ozone Hole, a Human-Caused Chemical Instability in the Stratosphere: What Should We Learn from It?" In *Geosphere-Biosphere Interactions and Climate*, ed. Lennart O. Bengtsson and Claus U. Hammer, 1–11. Cambridge, U.K.: Cambridge University Press, 2001.

Cuffey, K. M., and F. Vimeux. "Covariation of Carbon Dioxide and Temperature from the Vostok Ice Core after Deuterium-Excess Correction." *Nature* 412 (August 2, 2001): 523–525.

Culver, Stephen J., and Peter F. Rawson, eds. *Biotic Response to Global Change: The Last 145 Million Years.* Cambridge, U.K.: Cambridge University Press, 2000.

Cunningham, Dennis. "New Video Documents Climate Change Impacts on [the] High Arctic." International Institute for Sustainable Development, November 9, 2000. www.iisd.org/casl/projects/inuitobs.htm.

Curran, Mark A. J., Tas D. van Ommen, Vin I. Morgan, Katrina L. Phillips, and Anne S. Palmer. "Ice Core Evidence for Antarctic Sea Ice Decline since the 1950s." *Science* 302 (November 14, 2003): 1203–1206.

Curry, Ruth, Bob Dickson, and Igor Yashayaev. "A Change in the Freshwater Balance of the Atlantic Ocean over the Past Four Decades." *Nature* 426 (December 18, 2003): 826–829.

Dai, Aiguo Dai, Kevin E. Trenberth, and Taotao Qian. "A Global Dataset of Palmer Drought Severity Index for 1870–2002: Relationship with Soil Moisture and Effects of Surface Warming." *Journal of Hydrometeorology* 5 (6) (December 2004): 1117–1130.

Dalton, Alastair. "Ice Pack Clue to Climate-Change Effects." *The Scotsman,* October 18, 2001, 7.

Dauncey, Guy, and Patrick Mazza. *Stormy Weather: 101 Solutions to Global Climate Change.* Gabriola Island, BC: New Society Publishers, 2001.

Davidson, E. A., and A. I. Hirsch. "Carbon Cycle: Fertile Forest Experiments." *Nature* 411 (May 24, 2001): 431–433.

Davidson, Keay. "Film's Tale of Icy Disaster Leaves the Experts Cold." *San Francisco Chronicle,* June 1, 2004, E-1.

———. "Going to Depths for Evidence of Global Warming: Heating Trend in North Pacific Baffles Researchers." *San Francisco Chronicle,* March 1, 2004, A-4.

———. "Media Goofed on Antarctic Data: Global Warming Interpretation Irks Scientists." *San Francisco Chronicle,* February 4, 2002, A-8.

———. "Study Has New Evidence of Global Warming: Data, Taken 27 Years Apart, Shows Less Heat Escaping Earth Now." *San Francisco Chronicle,* March 15, 2001, A-2.

Davies, R., J. Cartwright, J. Pike, and C. Line. "Early Oligocene Initiation of North Atlantic Deep Water Formation." *Nature* 410 (April 19, 2001): 917–920.

Davis, Neil. *Permafrost: A Guide to Frozen Ground in Transition.* Fairbanks: University of Alaska Press, 2001.

Davis, Robert E., Paul C. Knappenberger, Patrick J. Michaels, and Wendy M. Novicoff. "Changing Heat-Related Mortality in the United States." *Environmental Health Perspectives,* July 23, 2003. doi:10.1289/ehp.6336. http://dx.doi.org.

Daynes, Byron W., and Glen Sussman. "The 'Greenless' Response to Global Warming." *Current History,* December 2005, 438–443.

Dayton, Leigh. "'Scary' Science Finds Earth Heating Up Twice as Fast as Thought." *Sydney Australian,* January 27, 2005, 3.

Dean, Cornelia. "Louisiana's Marshes Fight for Their Lives." *New York Times,* November 15, 2005. www.nytimes.com/2005/11/15/science/earth/15marsh.html.

de Angelis, Hernán, and Pedro Skvarca. "Glacier Surge after Ice Shelf Collapse." *Science* 299 (March 7, 2003): 1560–1562.

DeConto, Robert M., and David Pollard. "Rapid Cenozoic Glaciation of Antarctica Induced by Declining Atmospheric CO_2." *Nature* 421 (January 16, 2003): 245–249.

"Deforestation Could Push Amazon Rainforest to Its End." UniScience News Net, July 3, 2001. www.Unisci.com.

Del Genio, Anthony D. "The Dust Settles on Water Vapor Feedback." *Science* 296 (April 26, 2002): 665–666.

Del Giorgio, Paul A., and Carlos M. Duarte. "Respiration in the Open Ocean." *Nature* 420 (November 28, 2002): 379–384.

Derbyshire, David. "Baffled Bumble Bee Lured Out Early by Changing Climate." *London Daily Telegraph*, March 12, 2004, 15.

———. "Global Warming Fails to Boost Butterfly Visitors." *London Daily Telegraph*, November 1, 2001, 13.

———. "'Heatwave' in the Antarctic Halves Penguin Colony." *London Daily Telegraph*, May 10, 2001, 6.

Diaz, Henry F., ed. *Climate Variability and Change in High Elevation Regions: Past, Present and Future.* Norwell, MA: Kluwer Academic, 2003.

———, and Raymond S. Bradley. "Temperature Variations during the Last Century at High-Elevation Sites." *Climatic Change* 36 (1997): 253–279.

———, and Vera Markgraf, eds. *El Niño and the Southern Oscillation: Multiscale Variability and Global and Regional Impacts.* Cambridge, U.K.: Cambridge University Press, 2000.

Dickens, Gerald R. "Global Change: Hydrocarbon-Driven Warming." *Nature* 429 (June 3, 2004): 513–515.

———. "A Methane Trigger for Rapid Warming?" Review of *Methane Hydrates in Quaternary Climate Change: The Clathrate Gun Hypothesis* by James P. Kennett, Kevin G. Cannariato, Ingrid L. Hendy, and Richard J. Behl. *Science* 299 (February 14, 2003): 1017.

Dickerson, Russell R., John P. Burrows, Lenart Granat, Sergio G. Armin Hansel, James E. Johnson, Christian Neusuess, et al. "Overview: The Cruise of the Research Vessel Ronald H. Brown during the Indian Ocean Experiment (INDOEX), 1999." Unpublished draft www.meto.umd.edu/~russ/overv.htm.

Dickson, B., I. Yashayaev, J. Meincke, B. Turrell, S. Dye, and J. Holfort. "Rapid Freshening of the Deep North Atlantic Ocean over the Past Four Decades." *Nature* 416 (April 25, 2002): 832–836.

Dlugokencky, E. J., K. A. Masrie, P. M. Lang, and P. P. Tans. "Continuing Decline in the Growth Rate of the Atmospheric Methane Burden." *Nature* 393 (June 4, 1999): 447–450.

Dobson, G. M. B., and D. N. Harrison. "Measurement of the Amount of Ozone in the Earth's Atmosphere, and Its Relation to Other Geophysical Conditions." *Proceedings of the Royal Society* A110 (1926): 660–693.

"Does Oil Have a Future?" (Editorial). *Atlantic Monthly*, October 2005, 31–32.

Dominè, Florent, and Paul B. Shepson. "Air-Snow Interactions and Atmospheric Chemistry." *Science* 297 (August 30, 2002): 1506–1510.

Donn, Jeff. "New England's Brilliant Autumn Sugar Maples—and Their Syrup—Threatened by Warmth." Associated Press, September 23, 2002. (Lexis).

Doran, Peter T., John C. Priscu, W. Berry Lyons, John E. Walsh, Andrew G. Fountain, Diane M. McKnight, et al. "Antarctic Climate Cooling and Terrestrial Ecosystem Response." *Nature* 415 (January 30, 2002): 517–520.

Dore, John E., Roger Lukas, Daniel W. Sadler, and David M. Karl. "Climate-Driven Changes to the Atmospheric CO_2 Sink in the Subtropical North Pacific Ocean." *Nature* 424 (August 14, 2003): 754–757.

Drew, James. "Researchers at O.S.U. [Ohio State University] Have Global Change Signs on Ice." *Toledo (Ohio) Blade*, February 19, 2001.

"Drought, Excessive Heat Ruining Harvests in Western Europe." Associated Press in *Daytona Beach (Florida) News-Journal*, August 5, 2003, 3-A.

Droxler, Andrè W., Richard Z. Poore, and Lloyd H. Burckle, eds. *Earth's Climate and Orbital Eccentricity: The Marine Isotope Stage 11 Question.* Washington, D.C.: American Geophysical Union, 2003.

Drozdiak, William. "U.S. Firms Become 'Green' Advocates: Global Warming Talks Near End." *Washington Post*, November 24, 2000, E1.

Dube, Francine. "North America's Growing Season 12 Days Longer: 'What Is Good for Plants Is Not Necessarily Good for the Planet': Expert." *Canada National Post*, September 5, 2001. www.nationalpost.com.

Dunbar, Robert B. "El Niño: Clues from Corals." *Nature* 407 (October 26, 2000): 956–957.

Dunne, Nancy. "Climate Change Research Sparks Hawaii Protests." *London Financial Times*, June 20, 2002, 2.

"Dying Trees Release Air Polluting Chemical." Environment News Service, June 26, 2002. http://ens-news.com/ens/jun2002/2002-06-26-09.asp#anchor3.

Easterling, David R., Briony Horton, Phillip D. Jones, Thomas C. Peterson, Thomas R. Karl, David E. Parker, et al. "Maximum and Minimum Temperature Trends for the Globe." *Science* 277 (1997): 364–366.

———, Gerald A. Meehl, Camille Parmesan, Stanley A. Changnon, Thomas R. Karl, and Linda O. Mearns. "Climate Extremes: Observations, Modeling, and Impacts." *Science* 289 (September 22, 2000): 2068–2074.

Eckholm, Erik. "China Said to Sharply Reduce Emissions of Carbon Dioxide." *New York Times*, June 15, 2001, A-1.

EcoBridge. "What Can We Do about Global Warming?" No date. www
.ecobridge.org/content/g_wdo.htm.

Editorial. *Vancouver Province*, September 8, 2003, A-16.

"Editors' Choice: A Summary of Glaciation." *Science* 295 (January 18, 2002): 401.

Edwards, Martin, and Anthony J. Richardson. "Impact of Climate Change on
Maine Pelagic Phenology and Trophic Mismatch." *Nature* 430 (August 19,
2004): 881–884.

Egan, Timothy. "Alaska, No Longer So Frigid, Starts to Crack, Burn, and Sag."
New York Times, June 16, 2002, A-1.

———. "On Hot Trail of Tiny Killer in Alaska." *New York Times*, June 25, 2002, F-1.

Eggert, David. "Shareholders File Global Warming Resolutions at Ford, G[eneral]
M[otors]." Associated Press, December 11, 2002. (Lexis).

Eilperin, Juliet. "Warming Tied To Extinction of Frog Species." *Washington Post*,
January 12, 2006, A-1. www.washingtonpost.com/wp-dyn/content/article/
2006/01/11/AR2006011102121_pf.html.

Elderfield, Henry. "Climate Change: Carbonate Mysteries." *Science* 296 (May 31,
2002): 1618–1621.

Elliott, Valerie. "Polar Bears Surviving on Thin Ice." *London Times*, October 30,
2003. (Lexis).

Enger, Tim. "If This Is Spring, How Come I'm Still Shoveling Snow?" Letter to
the Editor. *Edmonton Journal*, May 8, 2003, A-19.

English, Andrew. "Feeding the Dragon: How Western Car-Makers Are Ignoring
Ecological Dangers in Their Rush to Exploit a Wide-Open Market." *London
Daily Telegraph*, October 30, 2004, 1.

English, Philip. "Ross Island Penguins Struggling." *New Zealand Herald*, January 9,
2003. (Lexis).

Environmental Research Foundation. "Profound Consequences: Climate Disruption, Contagious Disease, and Public Health." *Native Americas* 16 (3/4)
(Fall/Winter 1999): 64–67.

———. "Rachel's #466: Warming & Infectious Diseases," Annapolis, Maryland,
November 2, 1995. www.igc.apc.org/awea/wew/othersources/rachel466.html.

Epstein, Paul R. "Climate, Ecology, and Human Health." *Consequences* 3 (2) (1997):
3–19. www.gcrio.org/CONSEQUENCES/vol3no2/climhealth.html.

———, Henry F. Diaz, Scott Elias, Georg Grabherr, Nicohlas E. Graham, Willem
J. M. Martens, et al. "Biological and Physical Signs of Climate Change: Focus
on Mosquito-Borne Diseases." *Bulletin of the American Meteorological Society* 79
(3) (March 1998): 409–417.

Erbacher, Jochen, Brian T. Huber, Richard D. Norris, and Molly Markey. "Increased Thermohaline Stratification as a Possible Cause for an Ocean Anoxic
Event in the Cretaceous Period." *Nature* 409 (January 18, 2001): 325–327.

Erickson, Jim. "Boulder Team Sees Obstacle to Saving Ozone Layer: 'Rocks' in Arctic Clouds Hold Harmful Chemicals." *Rocky Mountain News*, February 9, 2001, 37-A.

———. "Glaciers Doff Their Ice Caps, and as Frozen Fields Melt, Anthropological Riches Are Revealed." *Rocky Mountain News*, August 22, 2002, 6-A.

———. "Going, Going, Gone? Front-Range Glaciers Declining: Researchers Point to a Warming World." *Rocky Mountain News*, October 26, 2004, 5-A.

Essex, Christopher, and Ross McKitrick. *Taken by Storm: The Troubled Science, Policy, and Politics of Global Warming.* Toronto: Kay Porter Books, 2002.

Etterson, Julie R., and Ruth G. Shaw. "Constraint to Adaptive Evolution in Response to Global Warming." *Science* 293 (October 5, 2001): 151–154.

"E.U. Plans to Become First Hydrogen Economy Superpower." *Industrial Environment* 12 (13) (December 2002). (Lexis).

"Europe's Heat Wave Toll Tops 19,000." Associated Press in *Omaha World-Herald*, September 26, 2003, 2.

Evans-Pritchard, Ambrose. "Dutch Have Only Years before Rising Seas Reclaim Land: Dikes No Match against Global Warming Effects." *London Daily Telegraph* in *Ottawa Citizen*, September 8, 2004, A-6.

"Expert Fears Warming Will Doom Bears." Canadian Press in *Victoria Times-Colonist*, January 5, 2003, C-8.

Fahey, D. W., R. S. Gao, K. S. Carslaw, J. Kettleborough, P. J. Popp, M. J. Northway, et al. "The Detection of Large HNO_3-Containing Particles in the Winter Arctic Stratosphere." *Science* 291 (February 9, 2001): 1026–1031.

Fahy, Declan. "Nature Charts Its Own Change: Irish Researchers Are Finding Signs of Climate Change in Trees and Bird Species." *Irish Times*, September 13, 2001. (Lexis).

Falkowski, P., R. J. Scholes, E. Boyle, J. Canadell, D. Canfield, J. Elser, et al. "The Global Carbon Cycle: A Test of Our Knowledge of Earth as a System." *Science* 290 (October 13, 2000): 291–296.

Fang, Jingyun, Anping Chen, Changhui Peng, Shuqing Zhao, and Longjun Ci. "Changes in Forest Biomass Carbon Storage in China between 1949 and 1998." *Science* 292 (June 22, 2001): 2320–2322.

Farman, J. C., B. G. Gardiner, and J. D. Shanklin. "Large Losses of Total Ozone Reveal Seasonal Clox/NOx Interaction." *Nature* 315 (1985): 207–210.

Faure, Michael, Joyetta Gupta, and Andries Nentjes, eds. *Climate Change and the Kyoto Protocol: The Role of Institutions and Instruments to Control Global Change.* Cheltenham, U.K.: Edward Edgar, 2003.

"Fear as Water Bleaches Reef." *Sydney (Australia) Daily Telegraph*, January 9, 2002, 12.

Feely, Richard A., Christopher L. Sabine, Kitack Lee, Will Berelson, Joanie Kleypas, Victoria J. Fabry, et al. "Impact of Anthropogenic CO_2 on the $CaCO_3$ System in the Oceans." *Science* 305 (July 16, 2004): 362–366.

Ferey, Marie-Pierre. "Floods, Droughts Loom from Climate Damage, Says Top Scientist." Agence France Presse, August 27, 2002. (Lexis).

Ferguson, H. L. "The Changing Atmosphere: Implications for Global Security." In *The Challenge of Global Warming*, ed. Dean Edwin Abrahamson, 48–62. Washington, D.C.: Island Press, 1989.

Fernandez, Art, Sheldon B. Davis, Jost O. L. Wendt, Roberta Cenni, R. Scott Young, and Mark L. Witten. "Particulate Emission from Biomass Combustion." *Nature* 409 (February 22, 2001): 998.

Fialka, John J. "Soot Storm: A Dirty Discovery over Indian Ocean Sets Off a Fight." *Wall Street Journal*, May 6, 2003, A-1, A-6.

Fidelman, Charlie. "Longer, Stronger Blazes Forecast." *Montreal Gazette*, July 11, 2002, A-4.

Field, Christopher B., and Michael R. Raupach, eds. *The Global Carbon Cycle: Integrating Humans, Climate, and the Natural World*. Washington, D.C.: Island Press, 2004.

Field, Michael. "Dying Pacific Breadfruit New Sign of Looming Disaster." Agence France Presse, December 1, 2002. (Lexis).

Finch, Gavin. "Falklands Penguins Dying in Thousands." *London Independent*, June 19, 2002, 12.

"Findings." *Washington Post*, April 30, 2002, A-30.

Finlayson-Pitts, Barbara J., and James N. Pitts Jr. *Chemistry of the Upper and Lower Atmosphere*. San Diego: Academic Press, 2000.

Finney, Bruce P., Irene Gregory-Eaves, Jon Sweetman, Marianne S. V. Douglas, and John P. Smol. "Impacts of Climatic Change and Fishing on Pacific Salmon Abundance over the Past 300 Years." *Science* 290 (October 27, 2000): 795–799.

Firor, John, and Judith Jacobsen. *The Crowded Greenhouse: Population, Climate Change, and Creating a Sustainable World*. New Haven, CT: Yale University Press, 2002.

"Fished to the Point of Ruin, North Sea Cod Stocks So Low as to Spell Disaster." *Glasgow Herald*, November 7, 2000, 18.

Fitter, A. H., and R. S. R. Fitter. "Rapid Changes in Flowering Time in British Plants." *Science* 296 (May 31, 2002): 1689–1691.

Flam, Faye. "It's Hot Now, but Scientists Predict There's an Ice Age Coming." *Philadelphia Inquirer*, August 23, 2002. (Lexis).

Fleck, John. "Dry Days, Warm Nights." *Albuquerque Journal*, December 28, 2003, B-1.

————. "Jack Frost's Nip Arrives a Bit Later." *Albuquerque Journal*, October 26, 2002, A-1.

Foley, Jonathan A. "Tipping Points in the Tundra." *Science* 310 (October 28, 2005): 627–628.

Fong, Petti. "Greenhouse Gas Chokes Sky after Wildfires." CanWest News Service in *Calgary Herald*, September 23, 2003, A-8.

Forero, Juan. "As Andean Glaciers Shrink, Water Worries Grow." *New York Times*, November 24, 2002, A-3.

Forest, Chris E., Peter H. Stone, Andrei P. Sokolov, Myles R. Allen, and Mort D. Webster. "Quantifying Uncertainties in Climate System Properties with the Use of Recent Climate Observations." *Science* 295 (January 4, 2002): 113–117.

Foster, Krishna L., Robert A. Plastridge, Jan W. Bottenheim, Paul B. Shepson, Barbara J. Finlayson-Pitts, and Chester W. Spicer. "The Role of Br_2 and BrCl in Surface Ozone Destruction at Polar Sunrise." *Science* 291 (January 19, 2001): 471–474.

Foukal, Peter, Gerald North, and Tom Wigley. "A Stellar View on Solar Variations and Climate." *Science* 306 (October 1, 2004): 68–69.

Fountain, Henry. "Observatory: Early Birds and Worms." *New York Times*, May 22, 2001, F-4.

————. "Observatory: Rice and Warm Weather." *New York Times*, June 29, 2004, F-1.

————. "Observatory: Threat to Rice Crops." *New York Times*, December 12, 2000, F-5.

Fowler, C., W. J. Emery, and J. Maslanik. "Satellite-Derived Evolution of Arctic Sea Ice Age: October 1978 to March 2003." *Geoscience and Remote Sensing Letters* 1 (2) (2004): 71–74.

Frappier, Amy, Dork Sahagian, Luis A. González, and Scott J. Carpenter. "El Niño Events Recorded by Stalagmite Carbon Isotopes." *Science* 298 (October 18, 2002): 565.

Freeman, C., C. D. Evans, and D. T. Monteith. "Export of Organic Carbon from Peat Soils." *Nature* 412 (August 23, 2001): 785.

————, N. Fenner, N. J. Ostle, H. Kang, D. J. Dowrick, B. Reynolds, et al. "Export of Dissolved Carbon from Peatlands under Elevated Carbon Dioxide Levels." *Nature* 430 (July 8, 2004): 195–198.

Freeman, James. "Methane Bubbles That Could Sink Ships." *Glasgow Herald*, November 2, 2000, 7.

————. "Ozone Repair Could Bring New Problem." *Glasgow Herald*, April 25, 2001, 3.

————, and Eleanor Cowie. "Pollutants Threaten the Great Barrier Reef." *Glasgow Herald*, January 25, 2002, 7.

Freemantle, Tony. "Global Warming Likely to Hit Texas: Scientists Say Temperature Rise Will Change Rainfall, Gulf Coast Region." *Houston Chronicle*, October 24, 2001, 32.

Frey, Darcy. "George Divoky's Planet." *New York Times Sunday Magazine*, January 6, 2002, 26–30.

Fried, Jeremy S., Margaret S. Torn, and Evan Mills. "The Impact of Climate Change on Wildfire Severity." *Climatic Change* 64 (May 2004): 169–191.

Fu, Quaing, Celeste M. Johansen, Stephen G. Warren, and Dian J. Seidel. "Contribution of Stratospheric Cooling to Satellite-Inferred Tropospheric Temperature Records." *Nature* 429 (May 6, 2004): 555–558.

Fukasawa, Masao, Howard Freeland, Ron Perkin, Tomowo Watanabe, Hiroshi Uchida, and Ayako Nishina. "Bottom Water Warming in the North Pacific Ocean." *Nature* 427 (February 26, 2004): 825–827.

Fung, Inez. "Variable Carbon Sinks." *Science* 290 (November 17, 2000): 1313.

Gaardner, Nancy. "State Enjoying 'Exceptional' Warmth." *Omaha World-Herald*, December 4, 2001, A-1, A-2.

Galloway, Elaine, and Chloe Rhodes. "Warm Spell Brings Early Start to Hay-Fever Misery." *London Evening Standard*, April 14, 2003, 16.

Ganeshram, Raja S. "Global Change: Oceanic Action at a Distance." *Nature* 419 (September 12, 2002): 123–125.

Ganopolski, Andrey, and Stefan Rahmstorf. "Rapid Changes of Glacial Climate Simulated in a Coupled Climate Model." *Nature* 409 (January 11, 2001): 153–158.

Gao, Yi Qin, and R. A. Marcus. "Strange and Unconventional Isotope Effects in Ozone Formation." *Science* 293 (July 13, 2001): 259–263.

Gardiner, Beth. "Report: Extreme Weather on the Rise, Likely to Get Worse." Associated Press Worldstream, International News, London, February 27, 2003. (Lexis).

Gardner, Toby A., Isabelle M. Côté, Jennifer A. Gill, Alastair Grant, and Andrew R. Watkinson. "Long-Term Region-Wide Declines in Caribbean Corals." *Science* 301 (August 15, 2003): 958–960. Posted online July 18, 2003, www.scienceexpress.org.

Gascard, J. C., A. J. Watson, M. J. Messias, K. A. Olsson, T. Johannessen, and K. Simonsen. "Long-Lived Vortices as a Mode of Deep Ventilation in the Greenland Sea." *Nature* 416 (April 4, 2002): 525–527.

Gasse, Françoise. "Hydrological Changes in Africa." *Science* 292 (June 22, 2001): 2259–2260.

———. "Kilimanjaro's Secrets Revealed." *Science* 298 (October 18, 2002): 548–549.

Gelbspan, Ross. "Boiling Point." *The Nation*, August 16, 2004, 24–27.

———. *Boiling Point: How Politicians, Big Oil and Coal, Journalists, and Activists Have Fueled the Climate Crisis—and What We Can Do to Avert Disaster.* New York: Basic Books (Perseus), 2004.

———. "A Global Warming." *The American Prospect* 31 (March/April 1997). www.prospect.org/archives/31/31gelbfs.html.

———. *The Heat Is On: The High-Stakes Battle over Earth's Threatened Climate.* Reading, MA: Addison-Wesley Publishing Co., 1997.

———. "The Heat Is On: The Warming of the World's Climate Sparks a Blaze of Denial." *Harper's*, December 1995. www.dieoff.com/page82.htm.

———, Gary Braasch, Mark Hertsgaard, Orna Izakson, David Helvarg, Jim Motavalli, et al. "Reality Check: The Global Warming Debate Is Over. It's Real, Inexorable, and Headed Our Way." *E: The Environmental Magazine*, September/October 2000, 24–39.

George, Jane. "Global Warming Threatens Nunavut's National Parks." *Nunatsiaq News*, May 19, 2000. www.nunatsiaq.com/archives/nunavut000531/nvt20519_18.html.

Gert-Jan Nabuurs, Gerd Folberth, Bernhard Schlamadinger, Ronald W. A. Hutjes, Reinhart Ceulemans, E.-Detlef Schulze, et al. "Europe's Terrestrial Biosphere Absorbs 7 to 12 Per Cent of European Anthropogenic CO_2 Emissions." *Science* 300 (June 6, 2003): 1538–1542.

Giannini, A., R. Saravanan, and P. Chang. "Oceanic Forcing of Sahel Rainfall on Interannual to Interdecadal Time Scales." *Science* 302 (November 7, 2003): 1027–1030.

"Giant Squid Film Team Makes Spectacular Catch." Agence France Presse, September 14, 2002.

Giardina, C., and M. Ryan. "Evidence That Decomposition Rates of Organic Carbon in Mineral Soil Do Not Vary with Temperature. *Nature* 404 (2000): 858–861.

Gill, R. A., H. W. Polley, H. B. Johnson, L. J. Anderson, H. Maherali, and R. B. Jackson. "Nonlinear Grassland Responses to Past and Future Atmospheric CO_2." *Nature* 417 (May 16, 2002): 279–282.

Gille, Sarah T. "Warming of the Southern Ocean since the 1950s." *Science* 295 (February 15, 2002): 1275–1277.

Gillett, Nathan P., and David W. J. Thompson. "Simulation of Recent Southern Hemisphere Climate Change." *Science* 302 (October 10, 2003): 273–275.

———, A. J. Weaver, F. W. Zwiers, and M. D. Flannigan. "Detecting the Effect of Climate Change on Canadian Forest Fires. *Geophysical Research Letters* 31, L18211. doi:10.1029/2004GL020876,2004.

Gillis, Justin. "A New Outlet for Venter's Energy: Genome Maverick to Take On Global Warming." *Washington Post*, April 30, 2002, E-1.

Gillon, Jim, and Dab Yakir. "Influence of Carbonic Anhydrase Activity in Terrestrial Vegetation on the Content of Atmospheric CO_2." *Science* 291 (March 30, 2001): 2584–2587.

Gjerdrum, Carina, Anne M. J. Vallee, Colleen Cassady St. Clair, Douglas F. Bertram, John L. Ryder, and Gwylim S. Blackburn. "Tufted Puffin Reproduction Reveals Ocean Climate Variability." *Proceedings of the National Academy of Sciences* 100 (16) (August 5, 2003): 9377–9382.

"Glacial Retreat Seen Worldwide." Environment News Service, May 30, 2002. http://ens-news.com/ens/may2002/2002-05-30-09.asp#anchor2; or www.gsfc.nasa.gov/topstory/20020530glaciers.html.

Glantz, Michael H. *Climate Affairs: A Primer.* Washington, D.C.: Island Press, 2003.

———. *Currents of Change: Impacts of El Niño and La Niña on Climate and Society,* 2nd ed. Cambridge, U.K.: Cambridge University Press, 2001.

Glick, Daniel. "The Heat Is On: Geosigns." *National Geographic,* September 2004, 12–33.

Glick, Patricia. *Global Warming: The High Costs of Inaction.* San Francisco: Sierra Club, 1998. www.sierraclub.org/global-warming/inaction.html.

"Global Climate Shift Feeds Spreading Deserts." Environment News Service, June 17, 2002. http://ens-news.com/ens/jun2002/2002-06-17-03.asp.

"Global Warming and Freak Winds Combine to Allow Explorers through Northeast Passage." *London Independent,* October 11, 2002, 14.

"Global Warming: An Insignificant Trend?" *Science* 292 (May 11, 2001): 1063.

"Global Warming Blamed for Rising Sea Levels." Associated Press in *Omaha World-Herald,* November 25, 2001, 20-A.

"Global Warming Could Hamper Ocean Sequestration." Environment News Service, December 4, 2002. http://ens-news.com/ens/dec2002/2002-12-04-09.asp.

"Global Warming Could Persist for Centuries." Environment News Service, February 18, 2002. http://ens-news.com/ens/feb2002/2002L-02-18-09.html.

"Global Warming Is Changing Tropical Forests." Environment News Service, August 7, 2002. http://ens-news.com/ens/aug2002/2002-08-07-01.asp.

"Global Warming Makes China's Glaciers Shrink by Equivalent of Yellow River." Agence France Presse, August 23, 2004. (Lexis).

"Global Warming Means More Snow for Great Lakes Region." AScribe Newswire, November 4, 2003. (Lexis).

"Global Warming Possible Factor in Mild Winter." *Tokyo Daily Yomiuri,* March 22, 2002, 2.

"Global-Warming Signal from the Ocean." *Dallas Morning News* in *New Orleans Times-Picayune,* December 31, 2000, 4.

"Global Warming's Sooty Smokescreen Revealed." New Scientist.com, June 3, 2003. www.newscientist.com/news/news.jsp?id=ns99993798.

"Global Warming Threatens California Water Supplies." Environment News Service, June 4, 2002. http://ens-news.com/ens/jun2002/2002-06-04-09.asp#anchor3.

"Global Warming Troubles Qinghai-Tibet Railway Construction." Xinhua (Chinese News Agency), April 30, 2003. (Lexis).

"Global Warning on Climate." *Montreal Gazette*, February 27, 2001, B-2.

Gobeil, Charles, Robie W. Macdonald, John N. Smith, and Luc Beaudin. "Atlantic Water Flow Pathways Revealed by Lead Contamination in Arctic Basin Sediments." *Science* 252 (August 17, 2001): 1301–1304.

Goldenberg, Stanley B., Christopher W. Landsea, Alberto M. Mestas-Nuñez, and William M. Gray. "The Recent Increase in Atlantic Hurricane Activity: Causes and Implications." *Science* 293 (July 20, 2001): 474–479.

Goldman, Erica. "Even in the High Arctic, Nothing Is Permanent." *Science* 297 (August 30, 2002): 1493–1494.

Golomb, Dan S., Meyer Steinberg, and Klaus S. Lackner. "Issues of Carbon Sequestration." Letter to the Editor. *Science* 301 (September 5, 2003): 1325.

Goodale, Christine L., and Eric A. Davidson. "Carbon Cycle: Uncertain Sinks in the Shrubs." *Nature* 418 (August 8, 2002): 601.

Gordon, Susan. "U.S. Pacific Northwest Gets Reduced Supply of Snow, Climate Study Says." *Tacoma News-Tribune*, February 7, 2003. (Lexis).

Goulden, M. L., S. C. Wofsy, J. W. Harden, S. E. Trumbone, P. M. Crill, S. T. Gower, et al. "Sensitivity of Boreal Forest Carbon Dioxide to Soil Thaw." *Science* 279 (January 9, 1998): 214–217.

Grady, Denise. "Managing Planet Earth: On an Altered Planet, New Diseases Emerge as Old Ones Re-emerge." *New York Times*, August 20, 2002, F-2.

Graf, Hans-F. "The Complex Interaction of Aerosols and Clouds." *Science* 303 (February 27, 2004): 1309–1311.

Grant, Christine. "Swelling Seas Eating Away at Country's Monuments." *The Scotsman*, December 24, 2001, 5.

Grant, Paul M. "Hydrogen Lifts Off—with a Heavy Load: The Dream of Clean, Usable Energy Needs to Reflect Practical Reality." *Nature* 424 (July 10, 2003): 129–130.

"Great Barrier Reef Is Springing Back to Life." *Western Daily Press (Australia)*, December 18, 2002, 11.

"Great Lakes Water Levels Dropping: Lowest in 35 Years." Associated Press in *Canada National Post*, January 4, 2002. www.nationalpost.com/.

Greenaway, Norma. "Disaster Toll from Weather Up Tenfold: Droughts, Floods Need More Damage Control, Report Says." *Edmonton Journal*, February 28, 2003, A-5.

"Greens Want Global Warming Examined in Bushfire Inquiry." Australian Associated Press, January 21, 2003. (Lexis).

Gregg, Watson W., and Margarita E. Conkright. "Decadal Changes in Global Ocean Chlorophyll." *Geophysical Research Letters* 29 (15) (2002): 1–4. doi: 10.1029/2002GL014689.

Gregory, Angela. "Fear of Rising Seas Drives More Tuvaluans to New Zealand." *New Zealand Herald*, February 19, 2003. (Lexis).

Gregory, Jonathan M., Philippe Huybrechts, and Sarah C. B. Raper. "Threatened Loss of the Greenland Ice Sheet." *Nature* 428 (April 8, 2004): 616.

Gretchen, Vogel. "Will the Future Dawn in the North?" *Science* 305 (August 13, 2004): 966–967.

Gribben, John. *Hothouse Earth: The Greenhouse Effect and Gaia.* London: Bantam Press, 1990.

Griffies, Stephen M. *Fundamentals of Ocean Climate Models.* Princeton, NJ: Princeton University Press, 2004.

Griffin, James. *Global Climate Change: The Science, Economics, and Politics.* Cheltenham, U.K.: Edward Elgar, 2003.

Gruber, Nicolas, Charles D. Keeling, and Nicholas R. Bates. "Interannual Variability in the North Atlantic Ocean Carbon Sink." *Science* 298 (December 20, 2002): 2374–2378.

"Guardian Special Report: Global Warming." *London Guardian*, November 14, 2000. www.guardian.co.uk/globalwarming/story/0,7369,397255,00.html.

Gugliotta, Guy. "In Antarctica, No Warming Trend: Scientists Find Temperatures Have Gotten Colder in Past Two Decades." *Washington Post*, January 14, 2002, A-2.

———. "Warming May Threaten 37 Per Cent of Species by 2050." *Washington Post*, January 8, 2004, A-1. www.washingtonpost.com/wp-dyn/articles/A63153-2004Jan7.html.

Gupta, Joyeeta. *Our Simmering Planet: What to Do about Global Warming?* London: Zed Books, 2001.

Haigh, Joanna D. "Climate Variability and the Influence of the Sun." *Science* 294 (December 7, 2001): 2109–2111.

Haines, Andrew. "The Implications for Health." In *Global Warming: The Greenpeace Report*, ed. Jeremy Leggett, 149–162. New York: Oxford University Press, 1990.

Hakim, Danny. "Ford Executives Adopt Ambitious Plan to Rein in Global Warming." *New York Times* in *International Herald-Tribune*, October 5, 2004, 11.

———. "Ford Stresses Business, but Disappoints Environmentalists." *New York Times*, August 20, 2002, C-4.

———. "Several States Likely to Follow California on Car Emissions." *New York Times*, June 11, 2004, C-4.

Häkkinen, Sirpa, and Peter B. Rhines. "Decline of Subpolar North Atlantic Circulation during the 1990s." *Science* 304 (April 23, 2004): 555–559.

Hale, Ellen. "Seas Create Real Water Hazard: Changing Climate at Root of Erosion That's Putting Links Courses in Jeopardy." *USA Today*, July 18, 2001, 3-C.

"Half U.S. Climate Warming Due to Land Use Changes." Environment News Service, May 28, 2003. http://ens-news.com/ens/may2003/2003-05-28-01.asp.

Hall, Alan. "Postcards a Tell-Tale for Icy Retreat." *The Scotsman*, August 3, 2002, 8.

Hall, Alex, and Ronald J. Stouffer. "An Abrupt Climate Event in a Coupled Ocean-Atmosphere Simulation without External Forcing." *Nature* 409 (January 11, 2001): 171–174.

Hallam, Anthony, and Paul Wignall. *Mass Extinctions and Their Aftermath*. Oxford: Oxford University Press, 1997.

Hann, Judith. "Spring Wakes Early, but Will Autumn Lie in Late Again? What Will Tomorrow's World Look Like?" United Kingdom Woodland Trust, 2002. www.woodland-trust.org.uk/news/subindex.asp?aid=328.

"Hans Blix's Greatest Fear." *New York Times*, March 16, 2003, D-2.

Hansen, Bogi, Svein sterhus, Detlef Quadfasel, and William Turrell. "Already the Day after Tomorrow?" *Science* 305 (August 13, 2004): 953–954.

———, William R. Turrell, and Svein Sterhus. "Decreasing Overflow from the Nordic Seas into the Atlantic Ocean through the Faroe Bank Channel since 1950." *Nature* 411 (June 21, 2001): 927–930.

Hansen, James E. "Dangerous Anthropogenic Interference: A Discussion of Humanity's Faustian Climate Bargain and the Payments Coming Due." Paper presented at the Distinguished Public Lecture Series at the Department of Physics and Astronomy, University of Iowa, Iowa City, October 26, 2004.

———. "Defusing the Global Warming Time Bomb." *Scientific American* 290 (3) (March 2004): 68–77.

———. "The Greenhouse, the White House, and Our House." Typescript of a speech at the International Platform Association, Washington, D.C., August 3, 1989.

———. Is There Still Time to Avoid "Dangerous Anthropogenic Interference" with Global Climate? A Tribute to Charles David Keeling. Paper delivered to the American Geophysical Union, San Francisco, December 6, 2005. www.columbia.edu/~jeh1/keeling_talk_and_slides.pdf.

———. "An Open Letter on Global Warming." *Natural Science*, October 26, 2000. http://naturalscience.com/ns/letters/ns_let25.html.

————, D. Johnson, A. Lacis, S. Lebedeff, P. Lee, D. Rind, et al. "Climate Impact of Increasing Atmospheric Carbon Dioxide." *Science* 213 (August 28, 1981): 957–966.

————, and Larissa Nazarenko. "Soot Climate Forcing via Snow and Ice Albedos." *Proceedings of the National Academy of Sciences* 101 (2) (January 13, 2004): 423–428.

————, R. Ruedy, J. Glascoe, and M. Sato. "GISS Analysis of Surface Temperature Change." *Journal of Geophysical Research* 104 (December 27, 1999): 30997–31022.

————, R. Ruedy, A. Lacis, D. Koch, I. Tegen, T. Hall, et al. "Climate Forcings in Goddard Institute for Space Studies SI2000 Simulations." *Journal of Geophysical Research* 107 (2002): 4347. doi:10.1029/2001JD001143.

————, R. Ruedy, M. Sato, and K. Lo. "Global Warming Continues." *Science* 295 (January 11, 2002): 275.

————, and Makiko Sato. "Trends of Measured Climate Forcing Agents." *Proceedings of the National Academy of Sciences* 98 (December 18, 2001): 14778–14783.

Harries, J. E., H. E. Brindley, P. J. Sagoo, and R. J. Bantges. "Increases in Greenhouse Forcing Inferred from the Outgoing Long-Wave Radiation Spectra of the Earth in 1970 and 1997." *Nature* 410 (March 15, 2001): 355–357.

Hartmann, Dennis L. "Climate Change: Tropical Surprises." *Science* 295 (February 1, 2002): 811–812.

————, John M. Wallace, Varavut Limpasuvan, David W. J. Thompson, and James R. Holton. "Can Ozone Depletion and Global Warming Interact to Produce Rapid Climate Change?" *Proceedings of the National Academy of Sciences* 97 (4) (February 15, 2000): 1412–1417.

Harvell, C. Drew, Charles E. Mitchell, Jessica R. Ward, Sonia Altizer, Andrew P. Dobson, Richard S. Ostfeld, et al. "Climate Warming and Disease Risks for Terrestrial and Marine Biota." *Science* 296 (June 21, 2002): 2158–2162.

Harvey, Fiona. "Arctic May Have No Ice in Summer by 2070, Warns Climate Change Report." *London Financial Times*, November 2, 2004, 1.

Harvey, L. D. D. *Global Warming: The Hard Science*. Hoboken, NJ: Prentice-Hall, 2000.

Hasselmann, K., M. Latif, G. Hooss, C. Azar, O. Edenhofer, C. C. Jaeger, et al. "The Challenge of Long-Term Climate Change." *Science* 302 (December 12, 2003): 1923–1925.

Hassol, Susan Joy. *Impacts of a Warming Arctic: Arctic Climate Impact Assessment*. Cambridge, U.K.: Cambridge University Press, 2005.

Hatsuhisa, Takashima. "Climate." *Journal of Japanese Trade and Industry*, September 1, 2002. (Lexis).

Haug, Gerald H., Konrad A. Hughen, Daniel M. Sigman, Larry C. Peterson, and Ursula Röhl. "Southward Migration of the Intertropical Convergence Zone through the Holocene." *Science* 252 (August 17, 2001): 1304–1308.

Hay, S. I., J. Cox, D. J. Rogers, S. E. Randolph, D. I. Stern, G. D. Shanks, et al. "Climate Change and the Resurgence of Malaria in the East African Highlands." *Nature* 425 (February 21, 2002): 905–909.

———. "Climate Change (Communication Arising): Regional Warming and Malaria Resurgence." *Nature* 420 (December 12, 2002): 628.

Hayhoe, Katherine, Daniel Cayan, Christopher B. Field, Peter C. Frumhoff, Edwin P. Maurer, Norman L. Miller, et al. "Emissions Pathways, Climate Change, and Impacts on California." *Proceedings of the National Academy of Sciences* 101 (34) (August 24, 2004): 12422–12427.

Healy, Patrick. "Warming Waters: Lobstermen on Cape Cod Blame Light Hauls on Higher Ocean Temperatures." *Boston Globe*, August 30, 2002, B-1.

"Heavy Rains Threaten Flood-Prone Venice." *Singapore Straits Times*, June 8, 2002. (Lexis).

Heilprin, John. "Study Says Black Carbon Emissions in China and India Have Climate Change Effects." Associated Press, September 26, 2002. (Lexis).

———. "Study Sees Economic Benefits of Reducing Global Warming." Associated Press, July 13, 2001. www.worldwildlife.org/climate/climate.cfm.

Helmuth, Brian, Christopher D. G. Harley, Patricia M. Halpin, Michael O'Donnell, Gretchen E. Hofmann, and Carol A. Blanchette. "Climate Change and Latitudinal Patterns of Intertidal Thermal Stress." *Science* 298 (November 1, 2002): 1015–1017.

Henderson, Mark. "Antarctica Defies Global Warming." *London Times*, January 14, 2002.

———. "Hot News from 740,000 Years Ago Tells Us to Get Ready for Catastrophic Climate Change." *London Times*, June 10, 2004, 4.

———. "Past Ten Summers Were the Hottest in 500 Years." *London Times*, March 5, 2004, 10.

———. "Positive Winds Keeping Arctic Winters at Bay." *London Times*, July 6, 2001. (Lexis).

———. "Southern Krill Decline Threatens Whales, Seals." *London Times* in *Calgary Herald*, November 4, 2004, A-11.

———. " 'World Has 15 Years to Stop Global Warming.' " *London Times*, July 21, 2001. (Lexis).

Herbert, T. D., J. D. Schuffert, D. Andreasen, L. Heusser, M. Lyle, A. Mix, et al. "Collapse of the California Current during Glacial Maxima Linked to Climate Change on Land." *Science* 293 (July 6, 2001): 71–76.

Herrick, Thaddeus. "The New Texas Wind Rush: Oil Patch Turns to Turbines, as Ranchers Sell Wind Rights: A New Type of Prospector." *Wall Street Journal*, September 23, 2002, B-1, B-3.

Hertsgaard, Mark. "It's Much Too Late to Sweat Global Warming." *San Francisco Chronicle*, February 13, 2005.

Hesselbo, Stephen P., Darren R. Grocke, Hugh C. Jenkyns, Christian J. Bjerrum, Paul Farrimond, Helen S. Morgans Bell, et al. "Massive Dissociation of Gas Hydrate during a Jurassic Oceanic Anoxic Event." *Nature* 406 (July 27, 2000): 392–395.

Highfield, Roger. "Winter Floods 'Five Times More Likely.'" *London Daily Telegraph*, January 31, 2002, 8.

Hillaire-Marcel, C., A. de Vernal, G. Bilodeau, and A. J. Weaver. "Absence of Deep-Water Formation in the Labrador Sea during the Last Interglacial Period." *Nature* 410 (April 26, 2001): 1073–1077.

Hinrichs, Kai-Uwe, Laura R. Hmelo, and Sean P. Sylva. "Molecular Fossil Record of Elevated Methane Levels in Late Pleistocene Coastal Waters." *Science* 299 (February 21, 2003): 1214–1217.

Hintrichsen, Don. "The Oceans Are Coming Ashore." *World Watch*, November/December 2000, 26–35.

Hirsch, Jerry. "Damage to Coral Reefs Mounts, Study Says: Broad Survey Cites Human Causes such as Over-Fishing and Pollution; Reefs Are a Key Indicator of the Health of Oceans, Scientists Say." *Los Angeles Times*, August 26, 2002, 14.

Hodell, David A., Mark Brenner, Jason H. Curtis, and Thomas Guilderson. "Solar Forcing of Drought Frequency in the Maya Lowlands." *Science* 292 (May 18, 2001): 1367–1370.

Hodge, Amanda. "Patagonia's Big Melt 'Sign of Global Warming.'" *The (Sydney) Australian*, February 12, 2004, 8.

Hoegh-Guldberg, Ove, Ross J. Jones, Selina Ward, and William K. Loh. "Is Coral Bleaching Really Adaptive?" *Nature* 415 (February 7, 2001): 601–602.

Hoerling, Martin P., James W. Hurrell, and Taiyi Xu. "Tropical Origins for Recent North Atlantic Climate Change." *Science* 292 (April 6, 2001): 90–92.

———, and Arun Kumar. "The Perfect Ocean for Drought." *Science* 299 (January 31, 2003): 691–694.

Hoffert, Martin I., Ken Caldeira, Gregory Benford, David R. Criswell, Christopher Green, Howard Herzog, et al. "Advanced Technology Paths to Global Climate Stability: Energy for a Greenhouse Planet." *Science* 298 (November 1, 2002): 981–987.

Hoffheiser, Chuck. "Previous Hot Theory Was Global Cooling." Letter to the Editor. *Wall Street Journal*, June 19, 2001, A-23.

Hoffman, Ian. "Iron Curtain over Global Warming: Ocean Experiment Suggests Phytoplankton May Cool Climate." *Hayward (California) Daily Review*, April 17, 2004. (Lexis).

Hogan, Treacy. "Still Raining in Costa del Ireland." *Belfast Telegram*, November 26, 2002. (Lexis).

Hollingsworth, Jan. "Global Warming Studies Put Heat on State: Tampa Bay Area Labeled Extremely Vulnerable." *Tampa Tribune*, October 24, 2001, 1.

Holly, Chris. "Sea-Level Rise Seen as Key Global Warming Threat." *The Energy Daily* 32 (36) (February 25, 2004). (Lexis).

———. "World CO2 Emissions Up 13 Percent from 1990–2000." *Energy Daily* 30 (206) (October 25, 2002). (Lexis).

Hoskins, Brian J. "Climate Change at Cruising Altitude?" *Science* 301 (July 25, 2003): 469–470.

Hostetler, Steven W., and Peter U. Clark. "Tropical Climate at the Last Glacial Maximum Inferred from Glacier Mass-Balance Modeling." *Science* 290 (December 1, 2000): 1747–1750.

"Hot Polymer Catches Carbon Dioxide." Environment News Service, May 29, 2002. http://ens-news.com/ens/may2002/2002-05-29-05.asp.

"Hot Times in the City Getting Hotter." Environment News Service, September 27, 2002. http://ens-news.com/ens/sep2002/2002-09-27-09.asp#anchor8.

Houghton, John. *Global Warming: The Complete Briefing*. Cambridge, U.K.: Cambridge University Press, 1997.

Houghton, J. T., Y. Ding, D. J. Griggs, M. Noguer, P. J. van der Linden, X. Dai, et al. *Climate Change 2001: The Scientific Basis. Contribution of Working Group I to the Third Assessment Report of the Intergovernmental Panel on Climate Change.* Cambridge, U.K.: Cambridge University Press, 2001.

Houlder, Vanessa. "Faster Global Warming Predicted: Met Office Research Has 'Mind-Blowing' Implications." *London Financial Times*, November 9, 2000, 2.

———. "Rise Predicted in Aviation Carbon Dioxide Emissions." *London Financial Times*, December 16, 2002, 2.

———. "Royal Society Calls for Carbon Levy or Permits." *London Financial Times*, November 18, 2002, 4.

"How About That Weather? The Answer Is Blowing in the Wind, Rain, Snow...." *Washington Post*, January 26, 2005, C-16.

"How to Combat Global Warming: In the End, the Only Real Solution May Be New Energy Technologies." *Business Week*, August 16, 2004, 108.

Howat, I. M., I. Joughin, S. Tulaczyk, and S. Gogineni. "Rapid Retreat and Acceleration of Helheim Glacier, East Greenland." *Geophysical Research Letters* 32 (2005). L22502, doi:10.1029/2005GL024737.

Hu, Feng Sheng, Darrell Kaufman, Sumiko Yoneji, David Nelson, Aldo Shemesh, Yongsong Huang, et al. "Cyclic Variation and Solar Forcing of Holocene Climate in the Alaskan Subarctic." *Science* 301 (September 26, 2003): 1890–1893.

Huang, Y., F. A. Street-Perrott, S. E. Metcalfe, M. Brenner, M. Moreland, and K. H. Freeman. "Climate Change as the Dominant Control on Glacial-Interglacial Variations in C_3 and C_4 Plant Abundance." *Science* 293 (August 31, 2001): 1647–1651.

Huber, Matthew, and Rodrigo Caballero. "Eocene El Niño: Evidence for Robust Tropical Dynamics in the 'Hothouse.'" *Science* 299 (February 7, 2003): 877–881.

"Hudson Bay Ice-Free by 2050, Scientists Say." Canadian Broadcasting Corporation, March 14, 2001. http://north.cbc.ca/cgi-bin/templates/view.cgi?/news/2001/03/14/14hudsonice.

Hughen, Konrad A., Timothy I. Eglinton, Li Xu, and Matthew Makou. "Abrupt Tropical Vegetation Response to Rapid Climate Changes." *Science* 304 (June 25, 2004): 1955–1959.

———, John R. Southon, Scott J. Lehman, and Jonathan T. Overpeck. "Synchronous Radiocarbon and Climate Shifts during the Last Deglaciation." *Science* 290 (December 8, 2000): 1951–1954.

Hughes, T. P., A. H. Baird, D. R. Bellwood, M. Card, S. R. Connolly, C. Folke, et al. "Climate Change, Human Impacts, and the Resilience of Coral Reefs." *Science* 31 (August 15, 2003): 929–933.

Hulbe, Christina L. "Glaciology: How Ice Sheets Flow." *Science* 294 (December 14, 2001): 2300–2301.

Human, Katy. "Disappearing Arctic Ice Chills Scientists: A University of Colorado Expert on Ice Worries That the Massive Melting Will Trigger Dramatic Changes in the World's Weather." *Denver Post*, October 5, 2004, B-2.

Hume, Stephen. "A Risk We Can't Afford: The Summer of Fire and the Winter of the Deluge Should Prove to the Naysayers That If We Wait Too Long to React to Climate Change We'll Be in Grave Peril." *Vancouver Sun*, October 23, 2003, A-13.

Hungate, Bruce A., Jeffrey S. Dukes, M. Rebecca Shaw, Yiqi Luo, and Christopher B. Field. "Nitrogen and Climate Change." *Science* 302 (November 28, 2003): 1512–1513.

———, Peter D. Stiling, Paul Dijkstra, Dale W. Johnson, Michael E. Ketterer, Graham J. Hymus, et al. "CO_2 Elicits Long-Term Decline in Nitrogen Fixation." *Science* 304 (May 28, 2004): 1291.

Hunten, Donald M. "Clues to the Martian Atmosphere." *Science* 294 (November 30, 2001): 1843–1844.

Huq, Saleemul. "Climate Change and Bangladesh." *Science* 294 (November 23, 2001): 1617.

Hurrell, James W., Yochanan Kushnir, and Martin Visbeck. "The North Atlantic Oscillation." *Science* 291 (January 26, 2001): 603–605.

Huston, Michael A., James H. Brown, Andrew P. Allen, and James F. Gillooly. "Heat and Biodiversity." Letter to the Editor. *Science* 299 (January 24, 2003): 511.

"Hydrogen Leakage Could Expand Ozone Depletion." Environment News Service, June 13, 2003. http://ens-news.com/ens/jun2003/2003-06-13-09.asp.

Hymon, Steve. "Early Snowmelt Ignites Global Warming Worries: Scientists Have Known Rising Temperatures Could Deplete Water Sources, but Data Show It May Already Be Happening." *Los Angeles Times*, June 28, 2004, B-1.

"Ice a Scarce Commodity on Arctic Rinks: Global Warming Blamed for Shortened Hockey Season." *Financial Post (Canada)*, January 7, 2003, A-3.

Ingham, John. "Fears for the Earth. All Down to Us: Hotter by the Minute." *London Express*, September 2, 2003, 19.

———. "Stingers Thrive as the Country Gets Warmer: Invasion of the Scorpions." *London Express*, June 18, 2004, 40.

———, and Barry Keevins. "U.K. Set to Be 6 C. Warmer by 2099: London May Be Just Like Madrid; Threat of Tornadoes; Cities Face Flooding; The Three-Day Winter." *London Express*, February 24, 2003, 17.

"Inhofe Calls Global Warming Warnings a Hoax." Associated Press, Oklahoma State and Local Wire, July 29, 2003. (Lexis).

Inkley, D. B., M. G. Anderson, A. R. Blaustein, V. R. Burkett, B. Felzer, B. Griffith, et al. *Global Climate Change and Wildlife in North America.* Washington, D.C.: The Wildlife Society, 2004. www.nwf.org/news.

"Iron Link to CO_2 Reductions Weakened." Environment News Service, April 10, 2003. http://ens-news.com/ens/apr2003/2003-04-10-09.asp#anchor8.

"Is It Already Too Late to Stop Global Warming?" *Insurance Day*, July 14, 2004. (Lexis).

"Is the Crux Solar Flux?" *Science* 295 (January 4, 2002): 15.

Iven, Chris. "Heat Strangles Fish." *Syracuse Post-Standard*, July 9, 2002. www.syracuse.com/news/poststandard/index.ssf?/base/news-0/1026207306743 1.xml.

Ivins, Molly. "Ignoring Problem Works—for a While." *Charleston (West Virginia) Gazette*, June 28, 2003, 4-A.

Jablonski, L. M., X. Wang, and P. S. Curtis. "Plant Reproduction under Elevated CO_2 Conditions: A Meta-Analysis of Reports on 79 Crop and Wild Species." *New Phytologist* 156 (2002): 9–26.

Jackson, Derrick Z. "Sweltering in a Winter Wonderland." *Boston Globe*, December 5, 2001, A-23.

Jackson, Robert B., Jay L. Banner, Esteban G. Jobbagy, William T. Pockman, and Diana H. Wall. "Ecosystem Carbon Loss with Woody Plant Invasion of Grasslands." *Nature* 418 (August 8, 2002): 623–626.

Jacobs, S. S., C. F. Giulivi, and P. A. Mele. "Freshening of the Ross Sea during the Late 20th Century." *Science* 297 (July 19, 2002): 386–389.

Jacobson, Mark. "Strong Radiative Heating Due to the Mixing State of Black Carbon in Atmospheric Aerosols." *Nature* 409 (February 8, 2001): 695–697.

Jager, J., and H. L. Ferguson. *Climate Change: Science, Impacts, and Policy. Proceedings of the Second World Climate Conference.* Cambridge, U.K.: Cambridge University Press, 1991.

Jayaraman, K. S. "Climate Model under Fire as Rains Fail India." *Nature* 418 (August 15, 2002): 713.

———. "Monsoon Rains Start to Ease India's Drought." *Nature* 423 (June 12, 2003): 673.

Jenkyns, Hugh C., Astrid Forster, Stefan Schouten, and Jaap S. Sinninghe Damste. "High Temperatures in the Late Cretaceous Arctic Ocean." *Nature* 432 (December 16, 2004): 888–892.

Jensen, Elizabeth. "Activists Take 'The Day After' for a Spin." *Los Angeles Times*, May 26, 2004. (Lexis).

Jensen, Mari N. "Climate Change: Consensus on Ecological Impacts Remains Elusive." *Science* 299 (January 3, 2003): 38.

Jepson, Paul, James K. Jarvie, Kathy MacKinnon, and Kathryn A. Monk. "The End for Indonesia's Lowland Forests?" *Science* 292 (May 4, 2001): 859–861.

Johannessen, Ola M., Kirill Khvorostovsky, Martin W. Miles, and Leonid P. Bobylev. "Recent Ice-Sheet Growth in the Interior of Greenland." *Science* 310 (November 11, 2005): 1013–1016.

Johansen, Bruce E. "Arctic Heat Wave." *The Progressive*, October 2001, 18–20.

———. *The Global Warming Desk Reference.* Westport, CT: Greenwood, 2001.

John Paul II. *On Social Concern (Sollicitudo Rei Socialis).* No. 42. Washington, D.C.: United States Catholic Conference, 1988.

———. *On the Hundredth Anniversary of Rerum Novarum (Centesimus Annus).* No. 32. Washington, D.C.: United States Catholic Conference, 1991.

Johnson, Andrew. "Climate to Bring New Gardening Revolution: Hot Summers and Wet Winters Could Kill Our Best-Loved Plants." *London Independent*, May 12, 2002, 5.

Jomelli, V., V. P. Pech, C. Chochillon, and D. Brunstein. "Geomorphic Variations of Debris Flows and Recent Climatic Change in the French Alps." *Climatic Change* 64 (1/2) (May 2004): 77–102.

Jones, C. D., P. M. Cox, R. L. H. Essery, D. L. Roberts, and M. J. Woodage. "Strong Carbon Cycle Feedbacks in a Climate Model with Interactive CO_2 and Sulphate Aerosols." *Geophysical Research Letters* 30 (9) (2003). doi:o.1029/2003GL016,867.

Jones, George, and Charles Clover. "Blair Warns of Climate Catastrophe: 'Shocked' Prime Minister Puts Pressure on U.S. and Russia over Emissions." *London Daily Telegraph*, September 15, 2004, 2.

Jones, P. D., T. J. Osborn, and K. R. Briffa. "The Evolution of Climate over the Last Millennium." *Science* 292 (April 27, 2001): 662–667.

Jordan, Steve. "Wary Insurers Suspect Climate Change." *Omaha World-Herald*, September 1, 2005, 7-A.

Joughin, I., W. Abdalati, and M. Fahnestock. "Large Fluctuations in Speed on Greenland's Jakobshavn Isbrae Glacier." *Nature* 432 (December 2, 2004): 608–610.

Joughin, Ian, and Slawek Tulaczyk. "Positive Mass Balance of the Ross Ice Streams, West Antarctica." *Science* 295 (January 18, 2002): 476–480.

Jucks, K. W., and R. J. Salawitch. "Future Changes in Atmospheric Ozone." In *Atmospheric Science across the Stratosphere*, ed. David E. Siskind, Stephen D. Eckermann, and Michael E. Summers, 241–256. Washington, D.C.: American Geophysical Union, 2000.

"Jumbo Squid Has a Message for Us: Changing Global Patterns Are Going to Bring Different Species into Our Waters." *Victoria (British Columbia) Times-Colonist*, October 8, 2004, A-10.

Jurgensen, John. "The Weather End Game: The Climate-Change Disaster at the Heart of 'Day after Tomorrow' May Be Overplayed, but the Global-Warming Threat Is Real." *Hartford Courant*, May 27, 2004, D-1.

Kaiser, Jocelyn. "Climate Change: 17 National Academies Endorse Kyoto." *Science* 292 (May 18, 2001): 1275–1277.

———. "Ecological Society of America Meeting: Global Warming, Insects Take the Stage at Snowbird." *Science* 289 (September 22, 2000): 2031–2032.

———. "Glaciology: Warmer Ocean Could Threaten Antarctic Ice Shelves." *Science* 302 (October 31, 2003): 759.

———. "Lean Winters Hinder Birds' Summertime Breeding Efforts." *Science* 301 (August 22, 2003): 1033.

———. "Reproductive Failure Threatens Bird Colonies on North Sea Coast." *Science* 305 (August 20, 2004): 1090.

———. "Soaking Up Carbon in Forests and Fields." *Science* 290 (November 3, 2000): 922.

Kakutani, Michiko. "Beware! Tree-Huggers Plot Evil to Save World." *New York Times*, December 13, 2004, E-1.

Kalkstein, Laurence S. "Direct Impacts in Cities." *Lancet* 342 (December 4, 1993): 1397–1400.

Kalnay, Eugenia, and Ming Cai. "Impact of Urbanization and Land-Use Change on Climate." *Nature* 423 (May 29, 2003): 528–531.

Kambayashi, Takehiko. "World Weather Prompts New Look at Kyoto." *Washington Times*, September 5, 2003, A-17.

Kammen, Daniel M., Timothy E. Lipman, Amory B. Lovins, Peter A. Lehman, John M. Eiler, Tracey K. Tromp, et al. "Assessing the Future Hydrogen Economy." *Science* 302 (October 10, 2003): 226–229.

Kane, R. L., and D. W. South. "The Likely Roles of Fossil Fuels in the Next 15, 50, and 100 Years, with or without Active Controls on Greenhouse-Gas Emissions." In *Limiting Greenhouse Effects: Controlling Carbon-Dioxide Emissions. Report of the Dahlem Workshop on Limiting the Greenhouse Effect, Berlin, September 9–14, 1990*, ed. G. I. Pearman, 189–227. New York: John Wiley & Sons, 1991.

Kaplow, Larry. "Solar Water Heaters: Israel Sets Standard for Energy; Cutting Dependence: Jerusalem's Alternative Energy Use a Lesson for United States." *Atlanta Journal and Constitution*, August 5, 2001, 1-P.

Karl, Thomas R., P. D. Jones, R. W. Knight, G. Kukla, N. Plummer, V. Razuvayev, et al. "A New Perspective on Recent Global Warming: Asymmetric Trends of Daily Maximum and Minimum Temperature." *Bulletin of the American Meteorological Society* 74 (1993): 1007–1023.

———, Richard W. Knight, and Bruce Baker. "The Record-Breaking Global Temperatures of 1997 and 1998: Evidence for an Increase in the Rate of Global Warming." *Geophysical Research Letters* 27 (March 1, 2000): 719–722.

———, and Kevin E. Trenberth. "Modern Global Climate Change." *Science* 302 (December 5, 2003): 1719–1723.

Karnosky, David F., et al., eds. *The Impact of Carbon Dioxide and Other Greenhouse Gases on Forest Ecosystems. Report No. 3 of the IUFRO Task Force on Environmental Change*. Oxford, U.K.: CABI, Wallingford, 2001.

Karoly, David J. "Ozone and Climate Change." *Science* 302 (October 10, 2003): 236–237.

———, Karl Braganza, Peter A. Stott, Julie M. Arblaster, Gerald A. Meehl, Anthony J. Broccoli, et al. "Detection of a Human Influence on North American Climate." *Science* 302 (November 14, 2003): 1200–1203.

Kaser, Georg, Douglas R. Hardy, Thomas M. A. Org, Raymond S. Bradley, and Tharsis M. Hyera. "Modern Glacier Retreat on Kilimanjaro as Evidence of Climate Change: Observations and Facts." *International Journal of Climatology* 24 (2004): 329–339.

Kaufman, Yoram J., Didier Tanre, and Oliver Boucher. "A Satellite View of Aerosols in the Climate System." *Nature* 419 (September 12, 2002): 215–223.

Keeling, Charles D., and Timothy P. Whorf. "Atmospheric CO_2 Records from Sites in the SIO Air Sampling Network." In *Trends: A Compendium of Data on Global Change*. Oak Ridge, TN: Carbon Dioxide Information Analysis Center, Oak Ridge National Laboratory, U.S. Department of Energy, 2004.

———. "The 1,800-Year Oceanic Tidal Cycle: A Possible Cause of Rapid Climate Change." *Proceedings of the National Academy of Sciences* 97 (8) (April 11, 2000): 3814–3819.

Kelleher, Lynne. "Look Who's Here: Tropical Fish Warming to Waters around Ireland." *London Sunday Mirror*, October 20, 2002, 15.

Kellogg, William W. "Theory of Climate Transition from Academic Challenge to Global Imperative." In *Greenhouse Glasnost: The Crisis of Global Warming*, ed. Terrell J. Minger, 99. New York: Ecco Press, 1990.

Kelly, Robert C. *The Carbon Conundrum: Global Warming and Energy Policy for the Third Millennium*. Houston, Texas: CountryWatch, 2002.

"Kelp Points to Worrying Sea Change." *Canberra Times*, August 30, 2004, A-8.

Kennedy, David. "Climate Change and Climate Science." *Science* 304 (June 11, 2004): 1565.

Kennedy, Donald. "The Hydrogen Solution." *Science* 305 (August 13, 2004): 917.

———. "New Climate News." *Science* 290 (November 10, 2000): 1091.

Kennett, James P., Kevin G. Cannariato, Ingrid L. Hendy, and Richard J. Behl. *Methane Hydrates in Quaternary Climate Change: The Clathrate Gun Hypothesis*. Washington, D.C.: American Geophysical Union, 2003.

Kerr, Richard A. "A Bit of Icy Antarctica Is Sliding toward the Sea." *Science* 305 (September 24, 2004): 1897.

———. "Can the Kyoto Climate Treaty Be Saved from Itself?" *Science* 290 (November 3, 2000): 920–921.

———. "Climate Change: A Little Sharper View of Global Warming." *Science* 294 (October 26, 2001): 765.

———. "Climate Change: More Science and a Carrot, Not a Stick." *Science* 295 (February 22, 2002): 1439.

———. "Climate Change: Sea Change in the Atlantic." *Science* 303 (January 2, 2004): 35.

———. "Climate Change: Three Degrees of Consensus." *Science* 305 (August 13, 2004): 932–934.

———. "Climate Modelers See Scorching Future as a Real Possibility." *Science* 307 (January 28, 2005): 497.

———. "Does a Climate Clock Get a Noisy Boost?" *Science* 290 (October 27, 2000): 697–698.

———. "A Dripping Wet Early Mars Emerging from New Pictures." *Science* 290 (December 8, 2000): 1879–1880.

———. "An Early Start for Greenhouse Warming?" *Science* 303 (January 16, 2004): 306.

———. "European Climate: Mild Winters Mostly Hot Air, Not Gulf Stream." *Science* 297 (September 27, 2002): 2202.

———. "Getting Warmer, However You Measure It." *Science* 304 (May 7, 2004): 805–807.

———. "Global Change: Who Pushed Whom Out of the Last Ice Age?" *Science* 299 (March 14, 2003): 1645.

———. "Global Warming: Bush Backs Spending for a 'Global Problem.'" *Science* 292 (June 15, 2001): 1978.

———. "Global Warming: Draft Report Affirms Human Influence." *Science* 288 (April 28, 2000): 589–590.

———. "Greenhouse Warming Passes One More Test." *Science* 292 (April 13, 2001): 193.

———. "In Mass Extinction, Timing Is All." *Science* 305 (September 17, 2004): 1705.

———. "Is Katrina a Harbinger of Still More Powerful Hurricanes?" *Science* 309 (September 16, 2005): 1807.

———. "It's Official: Humans Are Behind Most of Global Warming." *Science* 291 (January 26, 2001): 566.

———. "Major Challenges for Bush's Climate Initiative." *Science* 293 (July 13, 2001): 199–201.

———. "Ozone Depletion: A Brighter Outlook for Good Ozone." *Science* 297 (September 6, 2002): 1623–1625.

———. "Paleoclimate: Tropical Pacific a Key to Deglaciation." *Science* 299 (January 10, 2003): 183–184.

———. "A Perfect Ocean for Four Years of Globe-Girdling Drought." *Science* 299 (January 31, 2003): 636.

———. "Reducing Uncertainties of Global Warming." *Science* 295 (January 4, 2002): 29–31.

———. "Rising Global Temperature, Rising Uncertainty." *Science* 292 (April 13, 2001): 192–194.

———. "Scary Arctic Ice Loss? Blame the Wind." *Science* 307 (January 14, 2005): 203.

———. "Second Thoughts on Skill of El Niño Predictions." *Science* 290 (October 13, 2000): 257–258.

———. "Signs of a Warm, Ice-Free Arctic." *Science* 305 (September 17, 2004): 1693.

———. "Signs of Success in Forecasting El Niño." *Science* 297 (July 26, 2002): 497.

———. "A Single Climate Mover for Antarctica." *Science* 296 (May 3, 2002): 825–826.

———. "A Slowing Cog in the North Atlantic Ocean's Climate Machine." *Science* 304 (April 16, 2004): 371–373.

———. "A Smoking Gun for an Ancient Methane Discharge." *Science* 286 (November 19, 1999): 1465.

———. "Stratospheric 'Rocks' May Bode Ill for Ozone." *Science* 291 (February 9, 2001): 962–963.

———. "Too Little, Too Late, at the Climate Talks." *Science* 290 (December 1, 2000): 1663.

———. "The Tropics Return to the Climate System." *Science* 292 (April 27, 2001): 660–661.

———. "A Variable Sun and the Maya Collapse." *Science* 292 (May 18, 2001): 1293.

———. "A Variable Sun Paces Millennial Climate." *Science* 294 (November 16, 2001): 1431–1432.

———. "Vicissitudes of Ancient Climate." *Science* 303 (January 16, 2004): 307.

———. "A Warmer Arctic Means Change for All." *Science* 297 (August 30, 2002): 1490–1493.

———. "Warming Indian Ocean Wringing Moisture from the Sahel." *Science* 302 (October 10, 2003): 210–211.

———. "Weather Forecasting: Getting a Handle on the North's 'El Niño.'" *Science* 294 (October 19, 2001): 494–495.

———. "A Well-Intentioned Cleanup Gets Mixed Reviews." *Science* 290 (November 3, 2000): 921.

———. "Whither Arctic Ice? Less of It, for Sure." *Science* 297 (August 30, 2002): 1491.

Keys, David. "Global Warming: Methane Threatens to Repeat Ice-Age Meltdown: 55 Million Years Ago, a Massive Blast of Gas Drove Up Earth's Temperature 7 C. And Another Explosion Is in the Cards, Say the Experts." *London Independent*, June 16, 2001. (Lexis).

Kiesecker, Joseph M., Andrew R. Blaustein, and Lisa K. Belden. "Complex Causes of Amphibian Population Declines." *Nature* 410 (April 5, 2001): 681–684.

King, David A. "Climate Change Science: Adapt, Mitigate, or Ignore?" *Science* 303 (January 9, 2004): 176–177.

Kininmonth, William. *Climate Change: A Natural Hazard.* Brentwood, U.K.: Multi-Science Publishing Co., 2004.

Kintisch, Eli. "Floods? Droughts? More Storms? Predictions Vary, but Scientists Agree: Our Climate Will Change; Midwesterners Could End Up Wet, or Dry." *St. Louis Post-Dispatch*, May 30, 2004, B-1.

Kirby, Alex. "Costing the Earth." British Broadcasting Corporation, Radio Four, October 26, 2000. http://news.bbc.co.uk/hi/english/sci/tech/newsid_990000/990391.stm.

Kirk-Davidoff, Daniel, Daniel P. Schrag, and James G. Anderson. "On the Feedback of Stratospheric Clouds on Polar Climate." *Geophysical Research Letters* 29 (11) (2002): 14659–14663.

Kitney, Geoff. "Global Warming Back On the Agenda." *Australian Financial Review*, May 29, 2004, 30.

Knapp, Alan K., Philip A. Fay, John M. Blair, Scott L. Collins, Melinda D. Smith, Jonathan D. Carlisle, et al. "Rainfall Variability, Carbon Cycling, and Plant Species Diversity in a Mesic Grassland." *Science* 298 (December 13, 2002): 2202–2205.

Knorr, Gregory, and Gerrit Lohmann. "Southern Ocean Origin for the Resumption of Atlantic Thermohaline Circulation during Deglaciation." *Nature* 424 (July 31, 2003): 532–536.

Knorr, W., I. C. Prentice, J. I. House, and E. A. Holland. "Long-Term Sensitivity of Soil Carbon Turnover to Warming." *Nature* 433 (January 20, 2005): 298–301.

Knutson, Thomas R., and Robert E. Tuleya. "Impact of CO_2-Induced Warming on Simulated Hurricane Intensity and Precipitation: Sensitivity to the Choice of Climate Model and Convective Parameterization." *Journal of Climate* 17 (18) (September 15, 2004): 3477–3495.

———, Robert E. Tuleya, and Yoshio Kurihara. "Simulated Increase of Hurricane Intensities in a CO_2-Warmed Climate." *Science* 297 (February 13, 1998): 1018–1020.

———, Robert E. Tuleya, Weixing Shen, and Isaac Ginis. "Impact of Co_2-Induced Warming on Hurricane Intensities as Simulated in a Hurricane Model with Ocean Coupling." *Journal of Climate* 14 (2001): 2458–2469.

Knutti, R., T. F. Stocker, F. Joos, and G. K. Plattner. "Constraints on Radiative Forcing and Future Climate Change from Observations and Climate Model Ensembles." *Nature* 416 (April 18, 2002): 719–723.

Koenig, Robert. "Climate Change: Experts Urge Speedup to Mine 'Archives.'" *Science* 293 (July 6, 2001): 31.

Kok, M. T. J., W. J. V. Vermeulen, A. P. C. Faaij, and D. de Jager, eds. *Global Warming and Social Innovation: The Challenge of a Climate-Neutral Society.* London: Earthscan, 2003.

Kolber, Zbigniew S., F. Gerald Plumley, Andrew S. Lang, J. Thomas Beatty, Robert E. Blankenship, Cindy L. VanDover, et al. "Contribution of Aerobic Photoheterotrophic Bacteria to the Carbon Cycle in the Ocean." *Science* 292 (June 29, 2001): 2492–2495.

Kondratyev, Kirill, Vladimir F. Krapivin, and Costas A. Varotsos. *Global Carbon Cycle and Climate Change*. Berlin, Germany: Springer/Praxis, 2003.

Konviser, Bruce I. "Glacier Lake Puts Global Warming on the Map." *Boston Globe*, July 16, 2002, C-1.

Koren, Ilan, Yoram J. Kaufman, Lorraine A. Remer, and Jose V. Martins. "Measurement of the Effects of Amazon Smoke on the Inhibition of Cloud Formation." *Science* 303 (February 27, 2004): 1342–1345.

Korner, Christian. "Slow In, Rapid Out—Carbon Flux Studies and Kyoto Targets." *Science* 300 (May 23, 2003): 1242–1243.

Koslov, Mikhail, and Natalia G. Berlina. "Decline in Length of the Summer Season on the Kola Peninsula, Russia." *Climatic Change* 54 (September 2002): 387–398.

Koutavas, Athanasios, Jean Lynch-Stieglitz, Thomas M. Marchitto Jr., and Julian P. Sachs. "El Niño-Like Pattern in Ice Age Tropical Pacific Sea Surface Temperature." *Science* 297 (July 12, 2002): 226–230.

Krajick, Kevin. "Arctic Life, on Thin Ice." *Science* 291 (January 19, 2001): 424–425.

———. "Ecology: Defending Deadwood." *Science* 293 (August 31, 2001): 1579–1581.

———. "Ice Man: Lonnie Thompson Scales the Peaks for Science." *Science* 298 (October 18, 2002): 518–522.

———. "Melting Glaciers Release Ancient Relics." *Science* 296 (April 19, 2002): 454–456.

———. "Tracing Icebergs for Clues to Climate Change." *Science* 292 (June 22, 2001): 2244–2245.

Kraus, Clifford, Steven Lee Myers, Andrew C. Revkin, and Simon Romero. "As Polar Ice Turns to Water, Dreams of Treasure Abound." *New York Times*, October 10, 2005. www.nytimes.com/2005/10/10/science/10arctic .html?ei=5094&en=64e93c8fc877d5f2&hp=&ex=1129003200&partner= homepage&pagewanted=print.

Krauthammer, Charles. "Blixful Amnesia." *Washington Post*, July 9, 2004, A-19.

Kristal-Schroder, Carrie. "British Adventurer's Polar Trek Foiled by Balmy Arctic." *Ottawa Citizen*, May 17, 2004, A-1.

Kristof, Nicholas. "Baked Alaska on the Menu?" *New York Times* in *Alameda (California) Times-Star*, September 14, 2003. (Lexis).

———. "The Storm Next Time." *New York Times*, September 1, 2005. www.nytimes.com.

Kump, Lee R. "Chill Taken Out of the Tropics." *Nature* 413 (October 4, 2001): 470–471.

———. "Reducing Uncertainty about Carbon Dioxide as a Climate Driver." *Nature* 419 (September 12, 2002): 188–190.

———. "What Drives Climate?" *Nature* 408 (December 7, 2000): 651–652.

Kuptana, Rosemarie. "Testimony of Rosemarie Kuptana, President, Inuit Circumpolar Conference." Second Conference of the Parties to the United Nations Framework Convention on Climate Change, Geneva, Switzerland, July 16–19, 1996.

Kuypers, Marcel M. M., Peter Blokker, Jochen Erbacher, Hanno Kinkel, Richard D. Pancost, Stefan Schouten, et al. "Massive Expansion of Marine Archaea during a Mid-Cretaceous Oceanic Anoxic Event." *Science* 293 (July 6, 2001): 92–95.

Labeyrie, Laurent. "Glacial Climate Instability." *Science* 290 (December 8, 2000): 1905–1907.

Lackner, Kalus S. "A Guide to CO_2 Sequestration." *Science* 300 (June 13, 2003): 1677–1678.

LaDeau, Shannon L., and James S. Clark. "Rising CO_2 Levels and the Fecundity of Forest Trees." *Science* 292 (April 6, 2001): 95–98.

Lai, R. "Soil Carbon Sequestration Impacts on Global Climate Change and Food Security." *Science* 304 (June 11, 2004): 1623–1627.

Lambeck, Kurt, Tezer M. Esat, and Emma Kate Potter. "Links between Climate and Sea Levels for the Past Three Million Years." *Nature* 419 (September 12, 2002): 199–206.

Lambert, S. "Changes in Winter Cyclone Frequencies and Strengths in Transient Enhanced Greenhouse Warming Simulations Using Two Coupled Climate Models." *Atmosphere-Ocean* 42 (3) (2004): 173–181.

Lamy, Frank, Jèrùme Kaiser, Ulysses Ninnemann, Dierk Hebbeln, Helge W. Arz, and Joseph Stoner. "Antarctic Timing of Surface Water Changes Off Chile and Patagonian Ice Sheet Response." *Science* 304 (June 25, 2004): 1959–1962.

Landauer, Robert. "Big Changes in Our China Suburb." *Sunday Oregonian*, October 20, 2002, F-4.

Landsea, Christopher. "NOAA: Report on Intensity of Tropical Cyclones." NOAA, Miami, Florida, August 12, 1999. www.aoml.noaa.gov/hrd/tcfaq/tcfaqG.html#G3.

Lane, Anthony. "The Current Cinema: Cold Comfort. *The Day after Tomorrow.*" *The New Yorker*, June 7, 2004, 102–103.

Langenberg, Heike. "Climate and Water." *Nature* 419 (September 12, 2002): 187.

"Late Snow Smoothes the Way for Iditarod Sled-Dog Race." Reuters in *Ottawa Citizen*, March 2, 2001, B-8.

Laurance, Jeremy. "Climate Change to Kill Thousands, Ministers Warned." *London Independent*, February 9, 2001, 2.

Laurance, William F., Mark A. Cochrane, Scott Bergen, Philip M. Fearnside, Patricia Delamônica, Christopher Barber, et al. "The Future of the Brazilian Amazon." *Science* 291 (January 19, 2001): 438–439.

———, Alexandre A. Oliveira, Susan G. Laurance, Richard Condit, Henrique E. M. Nascimento, Ana G. Sanchez-Torin, et al. "Pervasive Alteration of Tree Communities in Undisturbed Amazonian Forests." *Nature* 428 (March 11, 2004): 171–175.

Lavers, Chris. *Why Elephants Have Big Ears*. New York: St. Martin's Press, 2000.

Lawrence, David M., and Andrew G. Slater. "A Projection of Severe Near-surface Permafrost Degradation during the 21st Century." *Geophysical Research Letters* 32 (2005). L24401, doi:10.1029/2005GL025080.

Lawton, R. O., U. S. Nair, R. A. Pielke Sr., and R. M. Welch. "Climatic Impact of Tropical Lowland Deforestation on Nearby Montane Cloud Forests." *Science* 294 (October 19, 2001): 584–587.

Laxon, Seymour, Neil Peacock, and Doug Smith. "High Interannual Variability of Sea-Ice Thickness in the Arctic Region." *Nature* 425 (October 30, 2003): 947–950.

Lazaroff, Cat. "Aerosol Pollution Could Drain Earth's Water Cycle." Environment News Service, December 7, 2001. http://ens-news.com/ens/dec2001/2001L-12-07-06.html.

———. "Climate Change Threatens Global Biodiversity." Environment News Service, February 7, 2002. http://ens-news.com/ens/feb2002/2002L-02-07-06.html.

———. "Land Use Rivals Greenhouse Gases in Changing Climate." Environment News Service, October 2, 2002. http://ens-news.com/ens/oct2002/2002-10-02-06.asp.

———. "Loggerhead Turtle Sex Ratio Raises Concerns." Environment News Service, December 18, 2002. http://ens-news.com/ens/dec2002/2002-12-18-06.asp.

———. "Melting Arctic Permafrost May Accelerate Global Warming." Environment News Service, February 7, 2001. http://ens-news.com/ens/feb2001/2001L-02-07-06.html.

———. "Replacing Grass with Trees May Release Carbon." Environment News Service, August 8, 2002. http://ens-news.com/ens/aug2002/2002-08-08-07.asp.

Lea, David W. "The Glacial Tropical Pacific—Not Just a West Side Story." *Science* 297 (July 12, 2002): 202–203.

———. "Ice Ages, the California Current, and Devils Hole." *Science* 293 (July 6, 2001): 59–60.

———. "The 100,000-Year Cycle in Tropical SST [Sea-Surface Temperature], Greenhouse Forcing, and Climate Sensitivity." *Journal of Climate* 17 (11) (June 1, 2004): 2170–2179.

———, Dorothy K. Pak, Larry C. Peterson, and Konrad A. Hughen. "Synchroneity of Tropical and High-Latitude Atlantic Temperatures over the Last Glacial Termination." *Science* 301 (September 5, 2003): 1361–1364.

Leake, Jonathan. "Fiery Venus Used to Be Our Green Twin." *London Sunday Times*, December 15, 2002, 11.

Lean, Geoffrey. "Antarctica 'Melting before Our Eyes.' " *London Independent*, February 11, 2001, 10.

———. "Antarctic Becomes Too Hot for the Penguins: Decline of 'Dinner Jacket' Species Is a Warning to the World." *London Independent*, February 3, 2002, 9.

———. "The Big Thaw: Global Disaster Will Follow If the Ice Cap on Greenland Melts; Now Scientists Say It is Vanishing Far Faster than Even They Expected." *London Independent*, November 20, 2005. www.commondreams.org/headlines05/1120-03.htm.

———. "Experts Prove How Warming Changes World." *London Independent*, February 18, 2001, 12.

———. "Flooded Britain: If We Don't Act Now, It'll Be Too Late: Global Warming Causes Floods." *London Independent*, November 5, 2000, 16.

———. "Global Warming Will Redraw Map of the World." *London Independent*, November 7, 2004, 8.

———. "Hot Summer Sparks Global Food Crisis." *London Independent*, August 31, 2003, 4.

———. "Quarter of World's Corals Destroyed." *London Independent*, January 7, 2001, 7.

———. "U.K. Homes Face Huge New Threat from Floods." *London Independent*, September 15, 2002, 1.

———. "We Regret to Inform You That the Flight to Malaga Is Destroying the Planet: Air Travel Is Fast Becoming One of the Biggest Causes of Global Warming." *London Independent*, August 26, 2001, 23.

———. "Worst U.S. Drought in 500 Years Fuels Raging California Wildfires." *London Independent*, July 25, 2004, 20.

Lean, Judith, and David Rind. "Earth's Response to a Variable Sun." *Science* 292 (April 13, 2001): 234–236.

Lederer, Edith M. "U.N. Report Says Planet in Peril." Associated Press, August 13, 2002. (Lexis).

Leggett, Jeremy. *The Carbon War: Global Warming and the End of the Oil Era*. New York: Routledge, 2001.

————, ed. *Global Warming: The Greenpeace Report*. New York: Oxford University Press, 1990.

Leggett, Karby. "In Rural China, G[eneral] M[otors] Sees a Frugal but Huge Market: It Bets Tractor Substitute Will Look Pretty Good to Cold, Wet Farmers." *Wall Street Journal*, January 16, 2001, A-19.

Leidig, Michael, and Roya Nikkhah. "The Truth about Global Warming—It's the Sun That's to Blame: The Output of Solar Energy Is Higher Now Than for 1,000 Years, Scientists Calculate." *London Sunday Telegraph*, July 18, 2004, 5.

Lemke, Peter. "Open Windows to the Polar Oceans." *Science* 292 (June 1, 2001): 1670–1671.

Lemley, Brad. "The New Ice Age." *Discover* 23 (9) (September 2002): 35–41.

Levitus, Sydney, John I. Antonov, Julian Wang, Thomas L. Delworth, Keith W. Dixon, and Anthony J. Broccoli. "Anthropogenic Warming of Earth's Climate System." *Science* 292 (April 13, 2001): 267–270.

Lewis, Cynthia L., and Mary Alice Coffroth. "The Acquisition of Exogenous Algal Symbionts by an Octocoral after Bleaching." *Science* 304 (June 4, 2004): 1490–1492.

"Limiting Methane, Soot Could Quickly Curb Global Warming." Environment News Service, January 16, 2002. http://ens-news.com/ens/jan2002/2002L-01-16-01.html.

Linden, Eugene. "The Big Meltdown: As the Temperature Rises in the Arctic, It Sends a Chill around the Planet." *Time*, September 4, 2000. www.time.com/time/magazine/articles/0,3266,53418,00.html.

Linsley, Braddock K., Gerard M. Wellington, and Daniel P. Schrag. "Decadal Sea Surface Temperature Variability in the Subtropical South Pacific from 1726 to 1997 A.D." *Science* 290 (November 10, 2000): 1145–1148.

Lippert, John. "G[eneral] M[otors] Chief Weighs Future of Fuel Cells." *Toronto Star*, September 27, 2002, E-3.

"Liquid CO_2 Dump in Norwegian Sea Called Illegal." Environment News Service, July 11, 2002. http://ens-news.com/ens/jul2002/2002-07-11-02.asp.

Little, Angela F., Madeleine J. H. van Oppen, and Bette L. Willis. "Flexibility in Algal Endosymbioses Shapes Growth in Reef Corals." *Science* 304 (June 4, 2004): 1492–1494.

Liu, Yanggang, and Peter H. Daum. "Anthropogenic Aerosols: Indirect Warming Effect from Dispersion Forcing." *Nature* 419 (October 10, 2002): 580–581.

Livingstone, Daniel A. "Global Climate Change Strikes a Tropical Lake." *Science* 301 (July 25, 2003): 468–469.

Lloyd, Ian. "Global Warming: A $25-Billion Challenge?" *Science* 292 (May 11, 2001): 1063.

Loeb, V. "Effects of Sea-Ice Extent and Krill or Salp Dominance on the Antarctic Food Web." *Nature* 387 (1997): 897–900.

Lohmann, Ulrike, and Glen Lesins. "Stronger Constraints on the Anthropogenic Indirect Aerosol Effect." *Science* 298 (November 1, 2002): 1012–1015.

Lomborg, Bjorn. "Entertaining Discredited Ideas of a Climatic Catastrophe." *The Australian*, May 27, 2004. (Lexis).

Lovett, Richard A. "Global Warming: Rain Might Be Leading Carbon Sink Factor." *Science* 296 (June 7, 2002): 1787.

Lowy, Joan. "Effects of Climate Warming Are Here and Now." Scripps-Howard News Service, May 5, 2004. (Lexis).

Loya, Wendy M., and Paul Grogan. "Carbon Conundrum on the Tundra." *Nature* 431 (September 23, 2004): 406–407.

Lucht, Wolfgang, I. Colin Prentice, Ranga B. Myneni, Stephen Sitch, Pierre Friedlingstein, Wolfgang Cramer, et al. "Climatic Control of the High-Latitude Vegetation Greening Trend and Pinatubo Effect." *Science* 296 (May 31, 2002): 1687–1689.

Luhnow, David, and Geraldo Samor. "As Brazil Fills Up on Ethanol, it Weans off Energy Imports." *Wall Street Journal*, January 9, 2006, A-1, A-8.

Luterbacher, Jurg, Daniel Dietrich, Elena Xoplaki, Martin Grosjean, and Heinz Wanner. "European Seasonal and Annual Temperature Variability, Trends, and Extremes Since 1500." *Science* 303 (March 5, 2004): 1499–1503.

Lynas, Mark. *High Tide: News from a Warming World.* London: Flamingo, 2004.

———. *High Tide: The Truth about Our Climate Crisis.* New York: Picador/St. Martins, 2004.

———. "Meltdown: Alaska Is a Huge Oil Producer and Has Become Rich on the Proceeds. But It Has Suffered the Consequences: Global Warming, Faster and More Terrifyingly Than Anyone Could Have Predicted." *London Guardian* (Weekend Magazine), February 14, 2004, 22.

———. "Vanishing Worlds: A Family Snap[shot] of a Peruvian Glacier Sent Mark Lynas on a Journey of Discovery: With the Ravages of Global Warming, Would It Still Exist 20 Years Later?" *London Guardian*, March 31, 2004, 12.

Lynch-Stieglitz, Jean. "Hemispheric Asynchrony of Abrupt Climate Change." *Science* 304 (June 25, 2004): 1919–1920.

Macalister, Terry. "Confused Esso Tries to Silence Green Critics." *London Guardian*, June 25, 2002, 21.

———. "Shell Chief Delivers Global Warming Warning to Bush in His Own Back Yard." *London Guardian*, March 12, 2003, 19.

Macdougall, Doug. *Frozen Earth: The Once and Future Story of Ice Ages.* Berkeley: University of California Press, 2004.

Macey, Richard. "Climate Change Link to Clearing." *Sydney Morning Herald*, June 29, 2004, 2.

Mack, Michelle C., Edward A. G. Schuur, M. Syndonia Bret-Harte, Gaius R. Shaver, and F. Sturt Chapin III. "Ecosystem Carbon Storage in Arctic Tundra Reduced by Long-Term Nutrient Fertilization." *Nature* 432 (September 23, 2004): 440–443.

Macken, Julie. "The Big Dry: Bushfires Re-ignite Heated Debate on Global Warming." *Australian Financial Review*, February 17, 2003, 68.

———. "The Double-Whammy Drought." *Australian Financial Review*, May 4, 2004, 61.

Maher, B. A., and P. F. Dennis. "Evidence against Dust-Mediated Control of Glacial-Interglacial Changes in Atmospheric CO2." *Nature* 411 (May 10, 2001): 176–179.

"Maine Sets Global Warming Reduction Goals." Environment News Service, June 26, 2003. http://ens-news.com/ens/jun2003/2003-06-26-09.asp#anchor4.

"Major Temperature Rise Recorded in Arctic This Year: German Scientists." Agence France Presse, August 27, 2004. (Lexis).

Malakoff, David. "Deep-Sea Science: Cool Corals Become Hot Topic." *Science* 299 (January 10, 2003): 195.

———, and Richard Stone. "Scientists Recommend Ban on North Sea Cod." *Science* 298 (November 1, 2002): 939.

Malcolm, Jay R. *The Demise of an Ecosystem: Arctic Wildlife in a Changing Climate.* Washington, D.C.: World Wildlife Fund, 1996.

Malone, Thomas F., Edward D. Goldberg, and Walter H. Munk. Biographic Memoir for Roger Randall Dougan Revelle, 1909–1991. National Academy of Sciences. www.nap.edu/readingroom/books/biomems/rrevelle.html.

Mann, Michael E., Raymond S. Bradley, and Michael K. Hughes. "Global-Scale Temperature Patterns and Climate Forcing over the Past Six Centuries." *Nature* 392 (April 23, 1998): 779–787.

———, R. S. Bradley, and M. K. Hughes. "Northern Hemisphere Temperatures during the Past Millennium." *Geophysical Research Letters* 26 (1999): 759–762.

———, and Philip D. Jones. "Global Surface Temperatures over the Past Two Millennia." *Geophysical Research Letters* 30 (15) (August 2003). doi:10.1029/2003GL017814. www.ngdc.noaa.gov/paleo/pubs/mann2003b/mann2003b.html.

Manne, Alan S., and Richard G. Richels. "An Alternative Approach to Establishing Trade-Offs among Greenhouse Gases." *Nature* 410 (April 5, 2001): 675–677.

"Many U.S. Industry Giants Ignoring Global Warming." Environment News Service, July 9, 2003. http://ens-news.com/ens/jul2003/2003-07-09-11.asp.

Markandya, Anil, and Kirsten Halsnaes. *Climate Change and Sustainable Development: Prospects for Developing Countries.* London: Earthscan, 2002.

Marlow, Jeremy R., Carina B. Lange, Gerold Wefer, and Antoni Rosell-Melé. "Upwelling Intensification as Part of the Pliocene-Pleistocene Climate Transition." *Science* 290 (December 22, 2000): 2288–2291.

Marsh, Virginia. "Australia Expected to Become Hotter." *London Financial Times*, May 9, 2001, 18.

Martens, P., and A. J. McMichael, eds. *Environmental Change, Climate and Health Issues and Research Methods.* New York: Cambridge University Press, 2002.

Martens, Pim. "How Will Climate Change Affect Human Health?" *American Scientist* 87 (6) (November/December 1999): 534–541.

———, and Susanne C. Moser. "Health Impacts of Climate Change." *Science* 292 (May 11, 2001): 1063.

Martens, Willem J. M., Theo H. Jetten, and Dana A. Focks. "Sensitivity of Malaria, Schistosomiasis, and Dengue to Global Warming." *Climatic Change* 35 (1997): 145–156.

Maslanik, J. A., M. C. Serreze, and T. Agnew. "On the Record Reduction in 1998 Western Arctic Sea Ice Cover." *Geophysical Research Letters* 26 (13) (1999): 1905–1912.

Maslin, Mark. *Global Warming.* Stillwater, MN: Voyageur/Worldlife Library, 2002.

———, and Stephen J. Burns. "Reconstruction of the Amazon Basin Effective Moisture Availability over the Past 14,000 Years." *Science* 290 (December 22, 2000): 2285–2287.

Mason, John, Jack A. Bailey, and Ardea London. "Doomsday for Butterflies as Britain Warms Up: Dozens of Native Species at Risk of Extinction as Habitats Come under Threat." *London Independent*, September 29, 2002, 12.

Masson, Gordon. "Eco-Friendly Movement Growing in Music Biz." *Billboard*, March 15, 2003, 1.

Mastrandrea, Michael D., and Stephen H. Schneider. "Probabilistic Integrated Assessment of 'Dangerous' Climate Change." *Science* 304 (April 23, 2004): 571–575.

Maugh, Thomas H., II. "Global Warming Altering Mosquito." *Los Angeles Times*, November 12, 2001, A-18.

Maxwell, Fordyce. "Climate Warning for Scotland's Wildlife." *The Scotsman*, November 14, 2001, 7.

May, Wilhelm, Reinhard Voss, and Erich Roeckner. "Changes in the Mean and Extremes of the Hydrological Cycle in Europe under Enhanced Greenhouse

Gas Conditions in a Global Time-Slice Experiment." In *Climatic Change: Implications for the Hydrological Cycle and for Water Management*, ed. Martin Beniston, 1–30. Dordrecht, Germany: Kluwer Academic Publishers, 2002.

Mayewski, Paul Andrew, and Frank White. *The Ice Chronicles: The Quest to Understand Global Climate Change*. Hanover, NH: University Press of New England, 2002.

Mayle, Francis E., Rachel Burbridge, and Timothy J. Killeen. "Millennial-Scale Dynamics of Southern Amazonian Rain Forests." *Science* 290 (December 22, 2000): 2291–2294.

Maynard, Roger. "Climate Change Bringing More Floods to Australia." *Singapore Straits Times*, March 14, 2001, 17.

McCarthy, Michael. "Climate Change Provides Exotic Sea Life with a Warm Welcome to Britain." *London Independent*, January 24, 2002, 13.

———. "Climate Change Will Bankrupt the World." *London Independent*, November 24, 2000, 6.

———. "Countdown to Global Catastrophe." *London Independent*, January 24, 2005, 1.

———. "Ford Predicts End of Car Pollution: Boss Predicts the End of Petrol." *London Independent*, October 6, 2000, 10.

———. "The Four Degrees: How Europe's Hottest Summer Shows Global Warming Is Transforming Our World." *London Independent*, December 8, 2003, 3.

———. "Global Warming: Warm Spell Sees Nature Defying the Seasons: As 150 Countries Meet in Morocco to Discuss Climate Change, Britain's Natural World Responds to Record Temperatures." *London Independent*, October 30, 2001, 12.

———. "'Rainforests of the Sea' Ravaged: Over-Fishing and Pollution Kill 80 Per Cent of Coral on Caribbean Reefs." *London Independent*, July 18, 2003, 3.

———. "World's Largest Windmills Start Contributing to United Kingdom National Grid." *London Independent*, December 8, 2000, 9.

McCord, Joel. "Marshes in Decay Haunt the Bay." *Baltimore Sun*, December 6, 2000, 1-B.

McCrea, Steve. "Air Travel: Eco-Tourism's Hidden Pollution." *San Diego Earth Times*, August 1996. www.sdearthtimes.com/et0896/et0896s13.html.

McDermott, Frank, David P. Mattey, and Chris Hawkesworth. "Centennial-Scale Holocene Climate Variability Revealed by a High-Resolution...Record from Southwest Ireland." *Science* 294 (November 9, 2001): 1328–1331.

McElroy, Michael B. *The Atmospheric Environment: Effects of Human Activity*. Princeton, NJ: Princeton University Press, 2002.

McEwen, Bill. "The West's Dying Forests." Letter to the Editor. *New York Times*, August 2, 2004, A-16.

McFadden, Robert D. "New Orleans Begins a Search for Its Dead; Violence Persists." *New York Times*, September 5, 2005. https://unomail2.unomaha.edu/mail/uno/facstaff/bjohansen.nsf/38d46bf5e8f08834852564b500129b2c/5bcec3d2a6ada46a86257073002ee959?

McFarling, Usha Lee. "Fear Growing over a Sharp Climate Shift." *Los Angeles Times*, July 13, 2001, A-1.

———. "Glacial Melting Takes Human Toll: Avalanche in Russia and Other Disasters Show That Global Warming Is Beginning to Affect Areas Much Closer to Home." *Los Angeles Times*, September 25, 2002, A-4.

———. "NASA Finds 2002 Second Warmest Year on Record." *Los Angeles Times* in *Calgary Herald*, December 12, 2002, A-5.

———. "Scientists Now Fear 'Abrupt' Global Warming Changes: Severe and 'Unwelcome' Shifts Could Come in Decades, not Centuries, a National Academy Says in an Alert." *Los Angeles Times*, December 12, 2001, A-30.

———. "Shrinking Ice Cap Worries Scientists." *Los Angeles Times* in *Edmonton Journal*, December 8, 2002. www.canada.com/regina/story.asp?id={54910725-535A-4B0E-9A7E-FD7176D9C392}.

———. "Studies Point to Human Role in Global Warming." *Los Angeles Times*, April 13, 2001, A-1.

———. "Study Links Warming to Epidemics: The Survey Lists Species Hit by Outbreaks and Suggests That Humans Are Also in Peril." *Los Angeles Times*, June 21, 2002, A-7.

———. "A Tiny 'Early Warning' of Global Warming's Effect: The Population of Pikas, Rabbit-Like Mountain Dwellers, Is Falling, a Study Finds." *Los Angeles Times*, February 26, 2003, A-17.

———. "Warmer World Will Starve Many, Report Says." *Los Angeles Times*, July 11, 2001, A-3.

———, and Kenneth R. Weiss. "A Whale of a Food Shortage." *Los Angeles Times*, June 24, 2002, 1.

McKay, Paul. "Ford Leads Big Three in Green Makeover: Henry Ford Brought the World the Internal Combustion Engine. His Great Grandson May Consign It to History—and Help Preserve the Planet in the Bargain." *Ottawa Citizen*, May 28, 2001, A-1.

McKibben, Bill. "Bush in the Greenhouse." Review of *I.P.C.C. Third Assessment*. *New York Review of Books*, July 5, 2001, 35–38.

———. *The End of Nature*. New York: Random House, 1989.

———. "Worried? Us?" *Granta* 83 (Fall 2003): 7–12.

McKie, Robin. "Decades of Devastation Ahead as Global Warming Melts the Alps: A Mountain of Trouble as Matterhorn Is Rocked by Avalanches." *London Observer*, July 20, 2003, 18.

———. "Dying Seas Threaten Several Species: Global Warming Could Be Tearing Apart the Delicate Marine Food Chain, Spelling Doom for Everything from Zooplankton to Dolphins." *London Observer*, December 2, 2001, 14.

McLean, Jim. "Icebergs Fuel Fears on Climate." *Glasgow Herald*, December 11, 2000, 8.

McManus, Jerry F. "A Great Grand-Daddy of Ice Cores." *Nature* 429 (June 10, 2004): 611–612.

———, R. Francois, J.-M. Gherardi, L. D. Keigwin, and S. Brown-Leger. "Collapse and Rapid Resumption of Atlantic Meridional Circulation Linked to Deglacial Climate Changes." *Nature* 428 (April 22, 2004): 834–837.

McMichael, A. J. *Planetary Overload: Global Environmental Change and the Health of the Human Species*. Cambridge: Cambridge University Press, 1993.

McNeil, Ben I., Richard J. Matear, Robert M. Key, John L. Bullister, and Jorge L. Sarmiento. "Anthropogenic CO_2 Uptake by the Ocean Based on the Global Chlorofluorocarbon Data Set." *Science* 299 (January 10, 2003): 235–239.

McNeill, J. R. *Something New under the Sun: An Environmental History of the Twentieth-Century World*. New York: W. W. Norton, 2000.

McPhaden, Michael J., and Dongxiao Zhang. "Slowdown of the Meridional Overturning Circulation in the Upper Pacific Ocean." *Nature* 415 (February 7, 2002): 603–608.

McWilliams, Brendan. "Study of Plants Confirms Global Warming." *Irish Times*, November 1, 2001, 26.

Meagher, John. "Look What the Changing Climate Dragged in . . ." *Irish Independent*, July 9, 2004. (Lexis).

Meehl, Gerald A., and Claudia Tebaldi. "More Intense, More Frequent, and Longer Lasting Heat Waves in the 21st Century." *Science* 305 (August 13, 2004): 994–997.

Meek, James. "Global Warming Gives Pests Taste for Life in London." *London Guardian*, October 8, 2002, 6.

———. "Tropical Travellers: A Seahorse in the Thames." *London Guardian*, December 12, 2000, 4.

———. "Wildflowers Study Gives Clear Evidence of Global Warming." *London Guardian*, May 31, 2002, 6.

Meier, Mark F., and Mark B. Dyurgerov. "Sea-Level Changes: How Alaska Affects the World." *Science* 297 (July 19, 2002): 350–351.

————, and John M. Wahr. "Sea Level Is Rising: Do We Know Why?" *Proceedings of the Natural Academy of Sciences* 99 (May 14, 2002): 6524–6526.

Melillo, J. M., P. A. Steudler, J. D. Aber, K. Newkirk, H. Lux, F. P. Bowles, et al. "Soil Warming and Carbon-Cycle Feedbacks to the Climate System." *Science* 298 (December 13, 2002): 2173–2176.

"Melting of Ice at the North Pole to Be Surveyed." Inuit Circumpolar Conference Memo, January 8, 2002. From the files of Sheila Watt-Cloutier and Christian Schultz-Lorentzen.

"Melting Planet: Species are Dying Out Faster Than We Have Dared Recognize, Scientists Will Warn This Week." *London Independent*, October 2, 2005. http://news.independent.co.uk/world/environment/article316604.ece.

Melvin, Don. "Storm over Wind Energy: Britain's Renewable Power Push Stirs Turbulent Debate." *Atlanta Journal-Constitution*, July 5, 2004, 8-A.

————. "There'll Always Be an England? Study of Global Warming Says Sea Is Winning." *Atlanta Journal-Constitution*, June 5, 2004, 3-A.

Mendelsohn, Robert, ed. *Global Warming and the American Economy*. Cheltenham, U.K.: Edward Elgar, 2001.

Menon, Surabi, James Hansen, Larissa Nazarenko, and Yunfeng Luo. "Climate Effects of Black Carbon Aerosols in China and India." *Science* 297 (September 27, 2002): 2250–2253.

Meuvret, Odile. "Global Warming Could Turn Siberia into Disaster Zone: Expert." Agence France Presse, October 2, 2003.

Michaels, Patrick J. "*Day after Tomorrow*: A Lot of Hot Air." *USA Today*, May 25, 2004, 21-A.

————. *Meltdown: The Predictable Distortion of Global Warming by Scientists, Politicians, and the Media*. Washington, D.C.: Cato Institute, 2004.

Middlebrook, Ann M., and Margaret A. Tolbert. *Stratospheric Ozone Depletion*. Sausalito, CA: University Science, 2001.

Mieszkowski, Katharine. "Bush: Global Warming Is Just Hot Air." Salon.com, September 10, 2004. (Lexis).

Miles, Paul. "Fiji's Coral Reefs Are Being Ruined by Bleaching." *London Daily Telegraph*, June 2, 2001, 4.

Miller, Laury, and Bruce C. Douglas. "Mass and Volume Contributions to Twentieth-Century Global Sea Level Rise." *Nature* 428 (March 25, 2004): 406–408.

Milly, P. C. D., R. T. Wetherald, K. A. Dunne, and T. L. Delworth. "Increasing Risk of Great Floods in a Changing Climate." *Nature* 415 (January 30, 2002): 514–517.

Milmo, Cahal, and Elizabeth Nash. "Fish Farms Push Atlantic Salmon towards Extinction." *London Independent*, June 1, 2001, 11.

Selected Bibliography

Minnis, Patrick, J. Kirk Ayres, Rabindra Palikonda, and Dung Phan. "Contrails, Cirrus Trends, and Climate." *Journal of Climate* (April 5, 2004): 1671–1685.

Mitchell, John G. "Down the Drain: The Incredible Shrinking Great Lakes." *National Geographic*, September 2002, 34–51.

Mitrovica, Jerry X., Mark E. Tamisiea, James L. Davis, and Glenn A. Milne. "Recent Mass Balance of Polar Ice Sheets Inferred from Patterns of Global Sea-Level Change." *Nature* 409 (February 22, 2001): 1026–1029.

Moberg, Anders, Dmitry M. Sonechkin, Karin Holmgren, Nina M. Datsenko, and Wibjörn Karlén. "Highly Variable Northern Hemisphere Temperatures Reconstructed from Low- and High-Resolution Proxy Data." *Nature* 433 (February 10, 2005): 613–617.

Monastersky, Richard. "The Long Goodbye: Alaska's Glaciers Appear to Be Disappearing before Our Eyes: Are They a Sign of Things to Come?" *New Scientist* (April 14, 2001): 30–32.

Monin, Eric, Andreas Indermuhle, Andre Dallenbach, Jacqueline Fluckiger, Bernard Stauffer, Thomas F. Stocker, et al. "Atmospheric CO_2 Concentrations over the Last Glacial Termination." *Science* 291 (January 5, 2001): 112–114.

"Monster Iceberg Heads into Antarctic Waters." Agence France Presse, October 22, 2002. (Lexis).

Montaigne, Fen. "The Heat Is On: Ecosigns." *National Geographic*, September 2004, 34–55.

Moore, G. W. K., Gerald Holdsworth, and Keith Alverson. "Climate Change in the North Pacific Region over the Past Three Centuries." *Nature* 420 (November 28, 2002): 401–403.

Moran, Tom. "Scientists Trace Global Warming in [Alaska's] Interior." Associated Press, Alaska State Wire, July 10, 2003. (Lexis).

"More Carbon Dioxide Could Reduce Crop Value." Environment News Service, October 3, 2002. http://ens-news.com/ens/oct2002/2002-10-03-09.asp#anchor2.

"More Evidence Found of Warming in the Alaskan Arctic." Associated Press in *Omaha World-Herald*, June 3, 2001, 15-A.

Morell, Virginia. "The Heat Is On: Timesigns." *National Geographic*, September 2004, 56–75.

"More Than 80 Per Cent of Spain's Pyrenean Glaciers Melted Last Century." Agence France Presse, September 29, 2004. (Lexis).

Morgan, Vin, Marc Delmotte, Tas van Ommen, Jean Jouzel, Jèrùme Chappellaz, Suenor Woon, et al. "Relative Timing of Deglacial Climate Events in Antarctica and Greenland." *Science* 297 (September 13, 2002): 1862–1864.

Moritz, Richard E., Cecilia M. Bitz, and Eric J. Steig. "Dynamics of Recent Climate Change in the Arctic." *Science* 297 (August 30, 2002): 1497–1502.

Morton, Oliver. "The Tarps of Kilimanjaro." *New York Times*, November 17, 2003. www.nytimes.com/2003/11/17/opinion/17MORT.html.

Moss, Stephen. "Casualties." *London Guardian*, April 26, 2001, 18.

"Most Serious Greenhouse Gas Is Increasing, International Study Finds." *Science Daily*, April 27, 2001. www.sciencedaily.com/releases/2001/04/010427071254.htm.

Moy, Christopher M., Geoffrey O. Seltzer, Donald T. Rodbell, and David M. Anderson. "Variability of El Niño/Southern Oscillation Activity at Millennial Time-Scales during the Holocene Epoch." *Nature* 420 (November14, 2002): 162–165.

Mudelsee, Mandred, Michael Borngen, Gerd Tetzlaff, and Uwe Grunewald. "No Upward Trends in the Occurrence of Extreme Floods in Central Europe." *Nature* 425 (September 11, 2003): 166–169.

Müller, R., P. J. Crutzen, J.-U. Grooß, C. Brühl, J. M. Russel III, H. Gernandt, et al. "Severe Chemical Ozone Loss in the Arctic during the Winter of 1995–96." *Nature* 389 (October 16, 1997): 709–712.

Mundil, Roland, Kenneth R. Ludwig, Ian Metcalfe, and Paul R. Renne. "Age and Timing of the Permian Mass Extinctions: U/Pb Dating of Closed-System Zircons." *Science* 305 (September 17, 2004): 1760–1763.

Munk, Walter. "Ocean Freshening, Sea Level Rising." *Science* 300 (June 27, 2003): 2041–2043.

Munro, Margaret. "Earth's 'Big Burp' Triggered Warming: Prehistoric Release of Methane a Cautionary Tale for Today." *Edmonton Journal*, June 3, 2004, A-10.

———. "Global Warming Affecting Squirrels' Genes, Study Finds: Research in Yukon: 'Phenomenal Change' Seen as Rodents Breeding Earlier." *Canada National Post*, February 12, 2003, A-2.

———. "Puffin Colony Threatened by Warming: A Few Degrees Can Be Devastating. Thousands of Triangle Island Chicks Die When Heat Drives Off Their Favoured Fish." *Montreal Gazette*, July 15, 2003, A-12.

Murphy, Dean. "Study Finds Climate Shift Threatens California." *New York Times*, August 17, 2004. www.nytimes.com/2004/08/17/national/17heat.html.

Murphy, Kim. "Front-Row Exposure to Global Warming: Engineers Say Alaskan Village Could Be Lost as Sea Encroaches." *Los Angeles Times*, July 8, 2001, A-1.

Muscheller, R., J. Beer, G. Wagner, and R. C. Finkel. "Changes in Deep-Water Formation during the Younger Dryas Event Inferred from the 10Be and 14C Records." *Nature* 408 (November 30, 2000): 567–570.

Myneni, R. B., J. Dong, C. J. Tucker, R. K. Kaufmann, P. E. Kauppi, J. Liski, et al. "A Large Carbon Sink in the Woody Biomass of Nothern Forests." *Proceedings of the National Academy of Sciences* 98 (December 18, 2001): 14784–14789.

Mysak, Lawrence A. "Patterns of Arctic Circulation." *Science* 252 (August 17, 2001): 1269–1270.

Naftz, David L., David D. Susong, Paul F. Schuster, L. DeWayne Cecil, Michael D. Dettinger, Robert L. Michel, et al. "Ice Core Evidence of Rapid Air Temperature Increases since 1960 in the Alpine Areas of the Wind River Range, Wyoming, United States." *Journal of Geophysical Research* 107 (D13) (July 9, 2002): 4171–4187.

Naik, Gautam, and Geraldo Samor. "Drought Spotlights Extent of Damage in Amazon Basin." *Wall Street Journal*, October 21, 2005, A-12.

Naish, T. R., K. J. Woolfe, P. J. Barnett, G. S. Wilson, C. Atkins, S. M. Bohaty, et al. "Orbitally Induced Oscillations in the East Antarctic Ice Sheet at the Oligocene/Miocene Boundary." *Nature* 413 (October 18, 2001): 719–723.

Nakagawa, Takeshi, Hiroyuki Kitagawa, Yoshinori Yasuda, Pavel E. Tarasov, Kotoba Nishida, Katsuya Gotanda, Yuki Sawai, and Yangtze River Civilization Program Members. "Asynchronous Climate Changes in the North Atlantic and Japan during the Last Termination." *Science* 299 (January 31, 2003): 688–691.

Nance, John J. *What Goes Up: The Global Assault on Our Atmosphere.* New York: William Morrow and Co., 1991.

Naqvi, S. W. A., D. A. Jayakumar, P. V. Narvekar, H. Naik, V. V. S. S. Sarma, W. D'Souza, et al. "Increased Marine Production of N_2O Due to Intensifying Anoxia on the Indian Continental Shelf." *Nature* 408 (November 16, 2000): 346–349.

Nash, J. Madeleine. *El Niño: Unlocking the Secrets of the Master Weather-Maker.* New York: Warner Books, 2002.

National Academy of Sciences. *Policy Implications of Greenhouse Warming.* Washington, D.C.: National Academy Press, 1991.

National Assessment Synthesis Team of the U.S. Global Change Research Program. *Climate Change Impacts on the United States: The Potential Consequences of Political Variability and Change, Overview Report.* Cambridge, U.K.: Cambridge University Press, 2000.

Nemani, Ramakrishna R., Charles D. Keeling, Hirofumi Hashimoto, William M. Jolly, Stephen C. Piper, Compton J. Tucker, et al. "Climate-Driven Increases in Global Terrestrial Net Primary Production from 1982 to 1999." *Science* 300 (June 6, 2003): 1560–1563.

Nepstad, D., D. McGrath, A. Alencar, A. C. Barros, G. Carvalho, M. Santilli, et al. "Frontier Governance in Amazonia." *Science* 295 (January 25, 2002): 629–631.

Nesmith, Jeff. "Antarctic Glacier Melt Increases Dramatically." *Atlanta Journal-Constitution*, September 22, 2004, 9-A.

———. "Dirty Snow Spurs Global Warming: Study Says Soot Blocks Reflection, Hurries Melting." *Atlanta Journal-Constitution*, December 23, 2003, 3-A.

———. "Is the Earth Too Hot? A New Study Says No, but Then It Was Funded by Big Oil Companies." *Atlanta Journal and Constitution* in *Hamilton (Ontario) Spectator*, May 30, 2003. (Lexis).

———. "Sewage Off Keys Cripples Coral: Bacteria Causes Deadly Disease." *Atlanta Journal and Constitution*, June 18, 2002, 3-A.

"New Climate Model Predicts Greater 21st Century Warming." AScribe Newswire, May 19, 2003. (Lexis).

Newell, Peter. *Climate for Change: Non-State Actors and the Global Politics of the Greenhouse.* Cambridge, U.K.: Cambridge University Press, 2000.

"New E. U. Law Aims to Double Green Energy by 2010." Reuters, July 5, 2001. (Lexis).

Newkirk, Margaret. "Lloyd's Chief Sees No Relief in Premiums: Insurance Firms Rebuild Reserves." *Atlanta Journal-Constitution*, October 21, 2003, 3-D.

"New Look at Satellite Data Supports Global Warming Trend." AScribe Newswire, April 30, 2003. (Lexis).

Newman, Paul A. "Preserving Earth's Stratosphere." *Mechanical Engineering*, October 1998. www.memagazine.org/backissues/october98/features/stratos/stratos.html.

———, J. F. Gleason, R. D. McPeters, and R. S. Stolarski. "Anomalously Low Ozone over the Arctic." *Geophysical Research Letters* 24 (22) (November 15, 1997): 2689–2692.

"New NASA Satellite Sensor and Field Experiment Shows Aerosols Cool the Surface but Warm the Atmosphere." National Aeronautics and Space Administration Public Information Release, August 15, 2001. http://earthobservatory.nasa.gov/Newsroom/MediaResources/Indian_Ocean_Experiment/indoex_release.html.

"New Research Links Global Warming to Wildfires across the West." *Los Angeles Times* in *Omaha World-Herald*, November 5, 2004, 11-A.

"New Research Reveals 50-Year Sustained Antarctic Ice Decline." Agence France Presse, November 14, 2003. (Lexis).

"New Study Shows Global Warming Trend Greater without El Niño and Volcanic Influences." Environmental Journalists' Bulletin Board, December 13, 2000. environmentaljournalists@egroups.com.

"New Wave of Bleaching Hits Coral Reefs Worldwide." Environment News Service, October 29, 2002. http://ens-news.com/ens/oct2002/2002-10-29-19.asp#anchor1.

"New York Lakes Fail to Freeze." Environment News Service, March 21, 2002. http://ens-news.com/ens/mar2002/2002L-03-21-09.html#anchor1.

Nicholls, Neville. "The Changing Nature of Australian Droughts." *Climatic Change* 63 (2004): 323–336.

Nickerson, Colin. "An Early Melting Hurts Seals, Hunters in Canada." *Boston Globe*, April 1, 2002, A-1.

Nisbit, E. G., and B. Ingham. "Methane Output from Natural and Quasinatural Sources: A Review of the Potential for Change and for Biotic and Abiotic Feedbacks." In *Biotic Feedbacks in the Global Climate System: Will the Warming Feed the Warming?* ed. George M. Woodwell and Fred T. MacKenzie, 188–218. New York: Oxford University Press, 1995.

Nordhaus, William D. "Global Warming Economics." *Science* 294 (November 9, 2001): 1283–1284.

———, and Joseph Boyer. *Warming the World: Economic Models of Global Warming.* Cambridge, MA: Massachusetts Institute of Technology Press, 2000.

Nordic Council of Ministers. *The Nordic Arctic Environment: Unspoiled, Exploited, Polluted?* Copenhagen, Denmark: Nordic Council of Ministers, 1996.

Normile, Dennis. "Warmer Waters More Deadly to Coral Reefs Than Pollution." *Science* 290 (October 27, 2000): 682–683.

"North Pole Had Sub-Tropical Seas Because of Global Warming." Agence France Presse, September 7, 2004. (Lexis).

"North Pole Mussels Point to Warming, Scientists Say." Reuters in *Canada National Post*, September 18, 2004, A-16.

Northrop, Michael. "Adapting to Warming: The United States Can Do It, but Europe Can't." *Washington Post*, December 16, 2002, A-25.

"Norway Says No to Controversial Plan to Store CO_2 on Ocean Floor." Agence France Presse, August 22, 2002. (Lexis).

Nosengo, Niccola. "Venice Floods: Save Our City!" *Nature* 424 (August 7, 2003): 608–609.

Nussbaum, Alex. "The Coming Tide: Rise in Sea Level Likely to Increase N.J. Floods." *Bergen County (New Jersey) Record*, September 4, 2002, A-1.

Nuttall, Nick. "Coral Reefs 'On the Edge of Disaster.'" *London Times*, October 25, 2000. (Lexis).

———. "Global Warming Boosts El Niño." *London Times*, October 26, 2000. (Lexis).

———. "Strange Visitor Traced to Africa." *London Times*, December 11, 2000. (Lexis).

Obersteiner, M., Ch. Azar, P. Kauppi, K. Mollersten, J. Moreira, S. Nilsson, et al. "Managing Climate Risk." *Science* 294 (October 26, 2001): 785.

O'Carroll, Cynthia. "NASA Blames Greenhouse Gases for Wintertime Warming." *UniSci*, April 24, 2001. http://unisci.com/stories/20012/0424011.htm.

"Ocean Temperatures Reach Record Highs." Associated Press, September 9, 2002. (Lexis).

O'Connell, Sanjida. "A Cool Calculation: The Frozen Desert of Antarctica May Return to Lush Forest." *London Guardian*, November 29, 2001, 10.

———. "Power to the People." *London Times*, May 20, 2002. (Lexis).

O'Driscoll, Patrick. "2005 is Warmest Year on Record for Northern Hemisphere, Scientists Say." *USA Today*, December 16, 2005, 2-A.

Oechel, W. C., S. J. Hastings, G. Vourlitis, M. Jenkins, G. Richers, and N. Gruike. "Recent Change of Arctic Tundra Ecosystems from a Net Carbon Dioxide Sink to a Source." *Nature* 361 (1993): 520–523.

———, G. L. Vourlitis, and S. J. Hastings. "Cold-Season CO_2 Emission from Arctic Soils." *Global Biogeochemical Cycles* 11 (1997): 163–172.

———, G. L. Vourlitis, S. J. Hastings, and S. A. Bochkarev. "Effects of Arctic CO_2 Flux over Two Decades: Effects of Climate Change at Barrow, Alaska." *Ecological Applications* 5 (1995): 846–855.

Ogle, Andy. "Squirrels Get Squirrelier Earlier: Climate Change to Blame. Breeding Season in Yukon Advances 18 Days in Decade." *Edmonton Journal* (Canwest News Service) in *Montreal Gazette*, February 12, 2003, A-12.

O'Harra, Doug. "Marine Parasite Infects Yukon River King Salmon: Fish Are Left Inedible; Scientists Study Overall Impacts." *Anchorage Daily News*, January 28, 2004, A-1.

Ohmura, Atsum, and Martin Wild. "Climate Change: Is the Hydrological Cycle Accelerating?" *Science* 298 (November 15, 2002): 1345–1346.

Olsen, Jan M. "Europe Is Warned of Changing Climate." Associated Press, August 19, 2004. (Lexis).

Olson, Jeremy. "Flash Flooding Closes I-80." *Omaha World-Herald*, July 7, 2002, A-1.

O'Malley, Brendan. "Global Warming Puts Rainforest at Risk." *Cairns (Australia) Courier-Mail*, July 24, 2003, 14.

"100 Still Missing in Russian Avalanche." *Los Angeles Times*, September 24, 2002, A-5.

O'Neill, Brian C., and Michael Oppenheimer. "Dangerous Climate Impacts and the Kyoto Protocol." *Science* 296 (June 14, 2002): 1971–1972.

O'Neill, Graeme. "The Heat Is On." *Sydney (Australia) Sunday Herald-Sun*, March 31, 2002.

"One in Five ExxonMobil Shareholders Want Climate Action." Environment News Service, May 28, 2003. http://ens-news.com/ens/may2003/2003-05-28-09.asp#anchor3.

Oppenheimer, M., and R. B. Alley. "The West Antarctic Ice Sheet and Long-Term Climate Policy: An Editorial Comment." *Climatic Change* 64 (1/2) (May 2004): 1–10.

Oppenheimer, Michael, and Robert H. Boyle. *Dead Heat: The Race against the Greenhouse Effect.* New York: Basic Books, 1990.

O'Reilly, Catherine M., Simone R. Alin, Pierre-Denis Plisnier, Andrew S. Cohen, and Brent A. McKee. "Climate Change Decreases Aquatic Ecosystem Productivity of Lake Tanganyika, Africa." *Nature* 424 (August 14, 2003): 766–768.

Oren, R., D. S. Ellsworth, K. H. Johnsen, N. Phillips, B. E. Ewers, C. Maier, et al. "Soil Fertility Limits Carbon Sequestration by Forest Ecosystems in a CO_2-Enriched Atmosphere." *Nature* 411 (May 24, 2001): 469–472.

Orr, James C., Victoria J. Fabry, Olivier Aumont, Laurent Bopp, Scott C. Doney, Richard A. Feely, Anand Gnanadesikan, Nicolas Gruber, Akio Ishida, Fortunat Joos, Robert M. Key, Keith Lindsay, Ernst Maier-Reimer, Richard Matear, Patrick Monfray, Anne Mouchet, Raymond G. Najjar, Gian-Kasper Plattner, Keith B. Rodgers, Christopher L. Sabine, Jorge L. Sarmiento, Reiner Schlitzer, Richard D. Slater, Ian J. Totterdell, Marie-France Weirig, Yasuhiro Yamanaka, and Andrew Yooi. "Anthropogeenic Ocean Acidification over the Twenty-first Century and Its Impact on Calcifying Organisms." *Nature* 437 (September 29, 2005): 681–686.

Osterkamp, T., and V. Romanovsky. "Permafrost Monitoring and Detection of Climate Change: Comments." *Permafrost and Periglacial Processes* 9 (1998): 87–89.

"Over 80 Per Cent of Indonesia's Coral Reefs under Threat." *Jakarta Post*, September 13, 2001. (Lexis).

Overpeck, J. T. "Warm Climate Surprises." *Science* 271 (March 29, 1996): 1820.

Overpeck, J. T., M. Strum, J. A. Francis, D. K. Perovich, M. C. Serreze, R. Benner, E. C. Carmack, F. S. Chapin III, S. C. Gerlach, L. C. Hamilton, L. D. Hinzman, M. Holland, H. P. Huntington, J. R. Key, A. H. Lloyd, G. M. MacDonald, J. McFadden, D. Noone, T. D. Prowse, P. Schlosser, and C. Vorosmarty. "Artic System on Trajectory to New, Seasonally Ice-Free State." *EOS: Transactions of the American Geophysical Union* 86 (34) (August 23, 2005): 309, 312.

Pacala, S. W., G. C. Hurtt, D. Baker, P. Peylin, R. A. Houghton, R. A. Birdsey, et al. "Consistent Land- and Atmosphere-Based U.S. Carbon Sink Estimates." *Science* 292 (June 22, 2001): 2316–2320.

Pacala, S., and R. Socolow. "Stabilization Wedges: Solving the Climate Problem for the Next 50 Years with Current Technologies." *Science* 305 (August 13, 2004): 968–972.

"Pacific Too Hot for Corals of World's Largest Reef." Environment News Service, May 23, 2002. http://ens-news.com/ens/may2002/2002-05-23-01.asp.

Page, Susan E., Florian Siegert, John O. Rieley, Hans-Dieter V. Boehm, Adi Jaya, and Suwido Limin. "The Amount of Carbon Released from Peat and Forest Fires in Indonesia during 1997." *Nature* 420 (November 7, 2002): 61–65.

Paige, David A. "Global Change on Mars?" *Science* 294 (December 7, 2001): 2107–2108.

Paillard, Didler. "Glacial Hiccups." *Nature* 409 (January 11, 2001): 147–148.

Pallé, E., P. R. Goode, P. Montañés-Rodríguez, and S. E. Koonin. "Changes in Earth's Reflectance over the Past Two Decades." *Science* 304 (May 2004): 1299–1301.

Palmer, M. R., and P. N. Pearson. "A 23,000-Year Record of Surface Water pH and PCO_2 in the Western Equatorial Pacific Ocean." *Science* 300 (April 18, 2003): 480–482.

Palmer, T. N., and J. Ralsanen. "Quantifying the Risk of Extreme Seasonal Precipitation Events in a Changing Climate." *Nature* 415 (January 30, 2002): 512–514.

Pandolfi, John M., Roger H. Bradbury, Enric Sala, Terence P. Hughes, Karen A. Bjorndal, Richard G. Cooke, et al. "Global Trajectories of the Long-Term Decline of Coral Reef Ecosystems." *Science* 301 (August 15, 2003): 955–958.

Paraskevas, Joe. "Glaciers in the Canadian Rockies Shrinking to Their Lowest Level in 10,000 Years." *Canada National Post*, December 4, 2003, A-8.

Parmesan, Camille, and Gary Yohe. "A Globally Coherent Fingerprint of Climate Change Impacts across Natural Systems." *Nature* 421 (January 2, 2003): 37–42.

Parsons, Michael L. *Global Warming: The Truth behind the Myth.* New York: Plenum Press/Insight, 1995.

Paterson, W. S. B., and N. Reeh. "Thinning of the Ice Sheet in Northwest Greenland over the Past Forty Years." *Nature* 414 (November 1, 2001): 60–62.

Patterson, Kathryn L., James W. Porter, Kim B. Ritchie, Shawn W. Polson, Erich Mueller, Esther C. Peters, et al. "The Etiology of White Pox, a Lethal Disease of the Caribbean Elkhorn Coral, *Acropora palmata*." *Proceedings of the National Academy of Sciences* 99 (13) (June 25, 2002): 8725–8730.

Patz, Jonathan A., Mike Hulme, Cynthia Rosenzweig, Timothy D. Mitchell, Richard A. Goldberg, Andrew K. Githeko, et al. "Climate Change (Communication Arising): Regional Warming and Malaria Resurgence." *Nature* 420 (December 12, 2002): 627–628.

Payette, S., A. Delwaide, M. Caccianiga, and M. Beachemin. 2004. "Accelerated Thawing of Subarctic Peatland Permafrost over the Past 50 Years." *Geophysical Research Letters* 31, L18208, doi:10.1029/2004GL020358.

Payne, A. J., A. Vieli, A. P. Shepherd, D. J. Wingham, and E. Rignot. "Recent Dramatic Thinning of Largest West Antarctic Ice Stream Triggered by Oceans." *Geophysical Research Letters* 31 (23) (December 9, 2004). L23401, doi:10.1029/2004GL021284.

Payne, Jonathan L., Daniel J. Lehrmann, Jiayong Wei, Michael J. Orchard, Daniel P. Schrag, and Andrew H. Knoll. "Large Perturbations of the Carbon Cycle during Recovery from the End-Permian Extinction." *Science* 305 (July 23, 2004): 506–509.

Pearce, Fred. "Failing Ocean Current Raises Fears of Mini Ice Age." New Scientist.com News Service, November 30, 2005. www.newscientist.com/article.ns?id=dn8398.

———. *Global Warming*. New York: DK Publishing, 2002.

———. "Ground-Breaking Solutions to Global Warming." *London Independent*, December 8, 2000, 8.

———. "Massive Peat Burn Is Speeding Climate Change." New Scientist.com, November 3, 2004. www.newscientist.com/news/news.jsp?id=ns99996613.

Pearson, Paul N., Peter Ditchfield, and Nicholas J. Shackleton. "Palaeoclimatology (Communication Arising): Tropical Temperatures in Greenhouse Episodes." *Nature* 419 (October 31, 2002): 898.

———, Peter W. Ditchfield, Joyce Singano, Katherine G. Harcourt-Brown, Christopher J. Nicholas, Richard K. Olsson, et al. "Warm Tropical Sea-Surface Temperatures in the Late Cretaceous and Eocene Epochs." *Nature* 413 (October 4, 2001): 481–487.

———, and Martin R. Palmer. "Atmospheric Carbon Dioxide Concentrations over the Past 60 Million Years." *Nature* 406 (August 17, 2000): 695–699.

Pecher, Ingo A. "Oceanography: Gas Hydrates on the Brink." *Nature* 420 (December 12, 2002): 622–623.

Pegg, J. R. "The Earth Is Melting, Arctic Native Leader Warns." Environment News Service, September 16, 2004.

———. "Plants Prospering from Climate Change." Environment News Service, June 5, 2003. http://ens-news.com/ens/jun2003/2003-06-06-10.asp.

Pelton, Tom. "New Maps Highlight Vanishing E. Shore Coast: Technology Provides a Stark Forecast of the Combined Effect of Rising Sea Level and Sinking Land along the Bay." *Baltimore Sun*, July 30, 2004, 1-A.

Peng, Shaobing, Jianliang Huang, John E. Sheehy, Rebecca C. Laza, Romeo M. Visperas, Xuhua Zhong, et al. "Rice Yields Decline with Higher Night

Temperature from Global Warming." *Proceedings of the National Academy* of *Sciences* 101 (27) (July 6, 2004): 9971–9975.

Pennell, William, and Tim Barnett, eds. "The Effects of Climate Change on Water Resources in the West." Special issue, *Climatic Change* 62 (1–3) (January and February 2004).

Penner, Joyce E., Leon D. Rotstayn, and Thomas J. Crowley. "Indirect Aerosol Forcing." *Science* 290 (October 20, 2000): 407.

Pennisi, Elizabeth. "Climate Change: Early Birds May Miss the Worms." *Science* 292 (March 30, 2001): 2532.

———. "Survey Confirms Coral Reefs Are in Peril." *Science* 297 (September 6, 2002): 1622–1623.

Percy, Kevin E., Caroline S. Awmack, Richard L. Lindroth, Mark E. Kubiske, Brian J. Kopper, J. G. Isebrands, et al. "Altered Performance of Forest Pests under Atmospheres Enriched by CO_2 and O_3." *Nature* 420 (November 28, 2002): 403–407.

Perlman, David. "Decline in Oceans' Phytoplankton Alarms Scientists: Experts Pondering Whether Reduction of Marine Plant Life Is Linked to Warming of the Seas." *San Francisco Chronicle*, October 6, 2003, A-6.

———. "Shrinking Glaciers Evidence of Global Warming: Differences Seen by Looking at Photos from 100 Years Ago." *San Francisco Chronicle*, December 17, 2004, A-18.

Peterson, A. T., M. A. Ortega-Huerta, J. Bartley, V. Sanchez-Cordero, J. Soberson, R. H. Buddemeier, et al. "Future Projections for Mexican Fauna under Global Climate Change Scenarios." *Nature* 416 (April 11, 2002): 626–629.

Peterson, Bruce J., Robert M. Holmes, James W. McClelland, Charles J. Vorosmarty, Richard B. Lammers, Alexander I. Shiklomanov, et al. "Increasing River Discharge to the Arctic Ocean." *Science* 298 (December 13, 2002): 2171–2173.

Peterson, Larry C., Gerald H. Haug, Konrad A. Hughen, and Ursula Röhl. "Rapid Changes in the Hydrologic Cycle of the Tropical Atlantic during the Last Glacial." *Science* 290 (December 8, 2000): 1947–1951.

Petit, Charles W. "Arctic Thaw." *U.S. News and World Report*, November 8, 2004, 66–69.

Petrillo, Lisa. "Turning the Tide in Venice." Copley News Service, April 28, 2003. (Lexis).

Phelan, Amanda. "Turning Up the Heat." *Sydney Sunday Telegraph*, August 18, 2002, 47.

"Phew, What a Scorcher—and It's Going to Get Worse." Agence France Presse, December 1, 2004. (Lexis).

Philander, George. "Why Global Warming Is Controversial." *Science* 294 (December 7, 2001): 2105–2106.

Pianin, Eric. "A Baltimore without Orioles? Study Says Global Warming May Rob Maryland, Other States of Their Official Birds." *Washington Post*, March 4, 2002, A-3.

———. "Greenhouse Gases Decrease: Experts Cite U.S. Economic Decline, Warm Winter." *Washington Post*, December 21, 2002, A-2.

———. "On Global Warming, States Act Locally: At Odds with Bush's Rejection of Mandatory Cuts, Governors and Legislatures Enact Curbs on Greenhouse Gases." *Washington Post*, November 11, 2002, A-3.

———. "Study Fuels Worry over Glacial Melting: Research Shows Alaskan Ice Mass Vanishing at Twice Rate Previously Estimated." *Washington Post*, July 19, 2002, A-14.

Pickrell, John. "Scientists Shower Climate Change Delegates with Paper." *Science* 293 (July 13, 2001): 200.

Pielke, Roger. "Land Use Changes and Climate Change." *Philosophical Transactions: Mathematical, Physical & Engineering Sciences* [Journal of The Royal Society of London] (August 2002).

Pielke, Roger A., Sr., and Thomas N. Chase. Comment on "Contributions of Anthropogenic and Natural Forcing to Recent Tropopause Height Changes." *Science* 303 (March 19, 2004): 1771.

Pierrehumbert, Raymond T. "High Levels of Atmospheric Carbon Dioxide Necessary for the Termination of Global Glaciation." *Nature* 429 (June 10, 2004): 646–649.

———. "The Hydrologic Cycle in Deep-Time Climate Problems." *Nature* 419 (September 12, 2002): 191–198.

Pilkey, Orrin H., and Andrew G. Cooper. "Society and Sea Level Rise." *Science* 303 (March 19, 2004): 1781–1782.

Pimm, Stuart L., Márcio Ayres, Andrew Balmford, George Branch, Katrina Brandon, Thomas Brooks, et al. "Can We Defy Nature's End?" *Science* 293 (September 21, 2001): 2207–2208.

Piotrowski, Jan A. "Earth Science: Glaciers at Work." *Nature* 424 (August 14, 2003): 737–738.

Pittman, Craig. "Global Warming Report Warns: Seas Will Rise." *St. Petersburg Times*, October 24, 2001, 3-B.

"Planting Northern Forests Would Increase Global Warming." Press Release, *New Scientist*, July 11, 2001. www.newscientist.com/news/news.jsp?id=ns99991003.

Poggioli, Sylvia. "Venice Struggling with Increased Flooding." National Public Radio, *Morning Edition*, November 29, 2002. (Lexis).

Pohl, Otto. "New Jellyfish Problem Means Jellyfish Are Not the Only Problem." *New York Times*, May 21, 2002, F-3.

Polakovic, Gary. "Airborne Soot Is Significant Factor in Global Warming, Study Says." *Los Angeles Times*, May 15, 2003, A-30.

———. "Deforestation Far Away Hurts Rain Forests, Study Says: Downing Trees on Costa Rica's Coastal Plains Inhibits Cloud Formation in Distant Peaks. 'It's Incredibly Ominous,' a Scientist Says." *Los Angeles Times*, October 19, 2001, A-1.

———. "Earth Losing Air-Cleansing Ability, Study Says: Worldwide Decline in a Molecule That Fights Pollution Is Found, but Experts Call the Losses Slight and Not Alarming." *Los Angeles Times*, May 4, 2001, A-1.

———. "States Taking the Initiative to Fight Global Warming: Unhappy with Bush's Policies, Local Officials Work to Slow Climate Change." *Los Angeles Times*, October 7, 2001, A-1.

Polyak, Victor J., and Yemane Asmerom. "Late Holocene Climate and Cultural Changes in the Southwestern United States." *Science* 293 (October 5, 2001): 148–151.

Pomerance, Rafe. "The Dangers from Climate Warming: A Public Awakening." In *The Challenge of Global Warming*, ed. Edwin Abrahamson, 259–269. Washington, D.C.: Island Press, 1989.

Post, Eric, and Mads C. Forchhammer. "Spatial Synchrony of Local Populations Has Increased in Association with the Recent Northern Hemisphere Climate Trend." *Proceedings of the National Academy of Sciences* 101 (June 22, 2004): 9286–9290.

Potter, Thomas D., and Bradley R. Colman, eds. *Handbook of Weather, Climate, and Water: Dynamics, Climate, Physical Meteorology, Weather Systems, and Measurements*. Hoboken, NJ: Wiley-Interscience, 2003.

Poulsen, Christopher J. "Paleoclimate: A Balmy Arctic." *Nature* 432 (December 16, 2004): 814–815.

Pounds, J. Alan. "Climate and Amphibian Decline." *Nature* 410 (April 5, 2001): 639–640.

———, and Robert Puschendorf. "Clouded Futures." *Nature* 427 (January 8, 2004): 107–108.

———, Martin R. Bustamante, Luis A. Coloma, Jamie A. Consuegra, Michael P. L. Fogden, Pru N. Foster, Enrique La Marca, Karen L. Masters, Andres Merino-Viteri, Robert Puschendorf, Santiago R. Ron, G. Arturo Sanchez-Azofeifa, Christopher J. Still, and Bruce E. Young. "Widespread Amphibian Extinctions from Epidemic Disease Driven by Global Warming." *Nature* 439 (January 12, 2006): 161–167.

Powell, Michael. "Northeast Seen Getting Balmier: Studies Forecast Altered Scenery, Coast." *Washington Post*, December 17, 2001, A-3.

"Power Games: Britain Was a World Leader in Wind Farm Technology. So Why Are All Our Windmills Made in Denmark?" *London Guardian*, July 16, 2001, 2.

Powlson, David. "Will Soil Amplify Climate Change?" *Nature* 433 (January 20, 2005): 204–205.

Prather, Michael J. "An Environmental Experiment with H_2?" *Science* 302 (October 24, 2003): 581–582.

Prigg, Mark. "Despite All the Heavy Rain, That Was the Hottest June for 28 Years." *London Evening Standard*, July 1, 2004, A-9.

Proffitt, Fiona. "Reproductive Failure Threatens Bird Colonies on North Sea Coast." *Science* 305 (August 20, 2004): 1090.

Prospero, Joseph M., and Peter J. Lamb. "African Droughts and Dust Transport to the Caribbean: Climate Change Implications." *Science* 302 (November 7, 2003): 1024–1027.

Prothero, Donald R., Linda C. Ivany, and Elizabeth A. Nesbitt, eds. *From Greenhouse to Icehouse: The Marine Eocene-Oligocene Transition*. New York: Columbia University Press, 2003.

Pugh, David. *Changing Sea Levels: Effects of Tides, Weather and Climate*. New York: Cambridge University Press, 2004.

Putkonen, J. K., and G. Roe. "Rain-on-Snow Events Impact Soil Temperatures and Affect Ungulate Survival." *Geophysical Research Letters* 30 (4) (2003): 1188–1192.

Quadfasel, Detlef. "Oceanography: The Atlantic Heat Conveyor Slows." *Nature* 438 (December 1, 2005): 565–566.

Quay, Paul. "Ups and Downs of CO_2 Uptake." *Science* 298 (December 20, 2002): 2344.

Quayle, Wendy C., Lloyd S. Peck, Helen Peat, J. C. Ellis-Evans, and P. Richard Harrigan. "Extreme Responses to Climate Change in Antarctic Lakes." *Science* 295 (January 25, 2002): 645.

"Quebec's Smoky Warning." Editorial. *Baltimore Sun*, July 9, 2002, 10-A.

Rabe, Barry G. *Statehouse and Greenhouse: The Emerging Politics of American Climate Change Policy*. Washington, D.C.: Brookings Institution Press, 2004.

Radford, Tim. "Antarctic Ice Cap Is Getting Thinner: Scientists' Worries That the South Polar Ice Sheet Is Melting May Be Confirmed by the Dramatic Retreat of the Region's Biggest Glacier." *London Guardian*, February 2, 2001, 9.

———. "As the World Gets Hotter, Will Britain Get Colder? Plunging Temperatures Feared after Scientists Find Gulf Stream Changes." *London Guardian*, June 21, 2001, 3.

———. "Coral Reefs Face Total Destruction within 50 Years." *London Guardian*, September 6, 2001, 9.

———. "85 Per Cent of Alaskan Glaciers Melting at 'Incredible Rate.' " *London Guardian*, July 19, 2002, 9.

———. "Global Warming Threatens Britain with Little Ice Age." *London Guardian*, September 7, 2001, 9.

———. "Scientists Discover the Harbinger of Drought: Subtle Temperature Changes in Tropical Seas May Trigger Northern Hemisphere's Long, Dry Spells." *London Guardian*, January 31, 2003, 18.

———. "Ten Key Coral Reefs Shelter Much of Sea Life: American Association Scientists Identify Vulnerable Marine 'Hot Spots' with the Richest Biodiversity on Earth." *London Guardian*, February 15, 2002, 12.

———. "2020: The Drowned World." *London Guardian*, September 11, 2004, 10.

———. "World May Be Warming Up Even Faster: Climate Scientists Warn New Forests Would Make Effects Worse." *London Guardian*, November 9, 2000, 10.

———. "World Sickens as Heat Rises: Infections in Wildlife Spread as Pests Thrive in Climate Change." *London Guardian*, June 21, 2002, 7.

Radowitz, John von. "Calmer Sun Could Counteract Global Warming." Press Association, October 5, 2003. (Lexis).

———. "Global Warming 'Smoking Gun' Found in the Oceans." Press Associated Ltd., February 18, 2005. (Lexis).

Rahmstorf, Stefan. "Ocean Circulation and Climate during the Past 120,000 Years." *Nature* 419 (September 12, 2002): 207–214.

———. "Thermohaline Circulation: The Current Climate." *Nature* 421 (February 13, 2003): 699.

"Rain, Rain Go Away." *London Times*, October 23, 2001. (Lexis).

Rajeev, K., and V. Ramanathan. "Direct Observations of Clear-Sky Aerosol Radiative Forcing from Space during the Indian Ocean Experiment." *Journal of Geophysical Research/Atmospheres* 106 (D15) (August 16, 2001): 17221.

Ralston, Greg. "Study Admits Arctic Danger." *Yukon News*, November 15, 1996. http://yukonweb.com/community/yukon-news/1996/nov15.htmld/#study.

Ramanathan, V. "Observed Increases in Greenhouse Gases and Predicted Climatic Changes." In *The Challenge of Global Warming*, ed. Edwin Abrahamson, 239–247. Washington, D.C.: Island Press, 1989.

———, P. J. Crutzen, J. T. Kiehl, and D. Rosenfeld. "Aerosols, Climate, and the Hydrological Cycle." *Science* 294 (December 7, 2001): 2119–2124.

Ranta, Esa, Per Lundberg, Veijo Kaitala, and Nils Chr. Stenseth. "On the Crest of a Population Wave." *Science* 298 (November 1, 2002): 973–974.

"Rare Sighting of Wasp North of Arctic Circle Puzzles Residents." Canadian Broadcasting Corporation, September 9, 2004. www.cbc.ca/story/science/national/2004/09/09/wasp040909.html.

Rau, G. H., K. Caldeira, Brad A. Seibel, and Patrick J. Walsh. "Minimizing Effects of CO_2 Storage in Oceans." *Science* 295 (January 11, 2002): 275.

Raven, Peter. "Environment: Why We Must Worry." *Science* 253 (August 31, 2001): 1598.

Ravindranath, H. H., and Jayant A. Sathaye. *Climate Change and Developing Countries.* Norwell, MA: Kluwer Academic, 2002.

Raymond, Charles F. "Glaciology: Ice Sheets on the Move." *Science* 298 (December 13, 2002): 2147–2148.

Raymond, Peter A., and Jonathan J. Cole. "Increase in the Export of Alkalinity from North America's Largest River." *Science* 301 (July 4, 2003): 88–91.

Reany, Patricia. " 'Millions Will Die' unless Climate Policies Change." Reuters, November 6, 1997. http://benetton.dkrz.de:3688/homepages/georg/kimo/0254.html.

Recar, Paul. "Study: Elements Can Stunt Plant Growth." Associated Press Online, December 5, 2002. (Lexis).

"Recent Warming of Arctic May Affect World-Wide Climate." National Aeronautics and Space Administration Press Release, October 23, 2003. www.gsfc.nasa.gov/topstory/2003/1023esuice.html.

Reed, Nicholas. "Mild Winter Stirs Wildlife to Early Thoughts of Love." *Vancouver Sun*, February 12, 2003, B-1.

Regalado, Antonio. "Skeptics on Warming Are Criticized." *Wall Street Journal*, July 31, 2003, A-3, A-4.

Reich, Peter B., Jean Knops, David Tilman, Joseph Craine, David Ellsworth, Mark Tjoelker, et al. "Plant Diversity Enhances Ecosystem Responses to Elevated CO2 and Nitrogen Deposition." *Nature* 410 (April 12, 2001): 809–812.

Reid, K., and J. P. Croxall. "Environmental Response of Upper Trophic-Level Predators Reveals a System Change in an Antarctic Marine Ecosystem." *Proceedings of the Royal Society of London* B268 (2001): 377–384.

Reid, T. R. "As White Cliffs Are Crumbling, British Want Someone to Blame." *Washington Post*, May 13, 2001, A-23.

Reilly, John, Peter H. Stone, Chris E. Forest, Mort D. Webster, Henry D. Jacoby, and Ronald G. Prinn. "Uncertainty and Climate Change Assessments." *Science* 293 (July 20, 2001): 430–433.

Remington, Robert. "Goodbye to Glaciers: Thanks to Global Warming, Mountains—the World's Water Towers—Are Losing Their Ice. As It

Disappears, so Does an Irreplaceable Source of Water." *Financial Post (Canada)*, September 6, 2002, A-19.

Rempel, A. W., E. D. Waddington, J. S. Wettlaufer, and M. F. Worster. "Possible Displacement of the Climate Signal in Ancient Ice by Premelting and Anomalous Diffusion." *Nature* 411 (May 31, 2001): 568–571.

Renewing the Earth: An Invitation to Reflection and Action on Environment in Light of Catholic Social Teaching. Washington, D.C.: United States Catholic Conference, n.d., 3.

"Research Casts Doubt on China's Pollution Claims." *Washington Post*, August 15, 2001, A-16. www.washingtonpost.com/wp-dyn/articles/A10645-2001Aug14.html.

Retallack, G. J. "A 300-Million-Year Record of Atmospheric Carbon Dioxide from Fossil Plant Cuticles." *Nature* 411 (May 17, 2001): 287–290.

Revelle, R., and H. E. Suess. "Carbon Dioxide Exchange between Atmosphere and Ocean and the Question of an Increase of Atmospheric CO_2 during the Past Decades." *Tellus* 9 (1957): 18–27.

Revkin, Andrew C. "Antarctic Glaciers Quicken Pace to Sea: Warming Is Cited." *New York Times*, September 24, 2004, A-24.

———. "Antarctic Test Raises Hope on a Global-Warming Gas." *New York Times*, October 12, 2000, A-18.

———. "Both Sides Now: New Way That Clouds May Cool." *New York Times*, June 19, 2001, F-4.

———. "Bush vs. the Laureates: How Science Becomes a Partisan." *New York Times*, October 19, 2004, F-1.

———. "A Chilling Effect on the Great Global Melt." *New York Times*, January 18, 2002, A-17.

———. "Climate Debate Gets Its Icon: Mt. Kilimanjaro." *New York Times*, March 23, 2004. NYTimes.com.

———. "Climate Talks Shift Focus to How to Deal with Changes." *New York Times*, November 3, 2002, A-8.

———. "Forecast for a Warmer World: Deluge and Drought." *New York Times*, August 28, 2002, A-10.

———. "Global Warming Is Expected to Raise Hurricane Intensity." *New York Times*, September 30, 2004, A-20.

———. "A Message in Eroding Glacial Ice: Humans Are Turning Up the Heat," *New York Times*, February 19, 2001, A-1.

———. "New Climate Model Highlights Arctic's Vulnerability." *New York Times*, October 31, 2005. www.nytimes.com/2005/10/31/science/earth/01warm_web.html.

————. "Planting New Forests Can't Match Saving Old Ones in Cutting Greenhouse Gases, Study Finds." *New York Times*, September 22, 2000, A-23.

————. "Scientists Say a Quest for Clean Energy Must Begin Now." *New York Times*, November 1, 2002, A-6.

————. "Study Finds a Decline in Natural Air Cleanser." *New York Times*, May 4, 2001. (Lexis).

————. "Study of Antarctic Points to Rising Sea Levels." *New York Times*, March 7, 2003, A-8.

————. "Two New Studies Tie Rise in Ocean Heat to Greenhouse Gases." *New York Times*, April 13, 2001, A-15.

————. "2004 Was Fourth-Warmest Year Ever Recorded." *New York Times*, February 10, 2005. www.nytimes.com/2005/02/10/science/10warm.html.

Rex, Markus, P. von der Gathen, Alfred Wegener, R. J. Salawitch, N. R. P. Harris, M. P. Chipperfield, et al. "Arctic Ozone Loss and Climate Change." *Geophysical Research Letters* 31 (March 10, 2004). www.eurekalert.org/pub_releases/2004-03/agu-ajho311004.php.

Reynolds, James. "Earth Is Heading for Mass Extinction in Just a Century." *The Scotsman*, June 18, 2003, 6.

Richardson, Anthony J., and David S. Schoeman."Climate Impact on Plankton Ecosystems in the Northeast Atlantic." *Science* 305 (September 10, 2004): 1609–1612.

Richardson, Franci. "Sharks Take the Bait: Experts: Sightings in Maine an 'Unusual Circumstance.' " *Boston Herald*, August 11, 2002, 3.

Richardson, Michael. "Indonesian Peat Fires Stoke Rise of Pollution." *International Herald-Tribune*, December 13, 2002, 5.

Richey, J. E., J. M. Melack, A. K. Aufdenkampe, V. M. Ballester, and L. L. Hess. "Outgassing from Amazonian Rivers and Wetlands as a Large Tropical Source of Atmospheric CO_2." *Nature* 416 (April 11, 2002): 617–620.

Ridgwell, Andy J., Martin J. Kennedy, and Ken Caldeira. "Carbonate Deposition, Climate Stability, and Neoproterozoic Ice Ages." *Science* 302 (October 31, 2003): 859–862.

Rifkin, Jeremy. *The Hydrogen Economy: The Creation of the World-Wide Energy Web and the Redistribution of Power on Earth.* New York: Tarcher (Putnam), 2002.

Rignot, Eric, G. Casassa, P. Gogineni, W. Krabill, A. Rivera, and R. Thomas. "Accelerated Ice Discharge from the Antarctic Peninsula Following the Collapse of Larsen B Ice Shelf." *Geophysical Research Letters* 31 (18) (September 22, 2004). doi:10.1029/2004GL020697.

————, and Stanley S. Jacobs. "Rapid Bottom Melting Widespread Near Antarctic Ice Sheet Grounding Lines." *Science* 296 (June 14, 2002): 2020–2023.

————, Andrès Rivera, and Gino Casassa. "Contribution of the Patagonia Ice-fields of South America to Sea Level Rise." *Science* 302 (October 17, 2003): 434–437.

————, and Robert H. Thomas. "Mass Balance of Polar Ice Sheets." *Science* 297 (August 30, 2002): 1502–1506.

Rigor, I. G., and J. M. Wallace. "Variations in the Age of Arctic Sea-Ice and Summer Sea-Ice Extent." *Geophysical Research Letters* 31 (2004). doi:10.1029/2004GL019492.

Rind, D. "The Sun's Role in Climate Variations." *Science* 296 (April 26, 2002): 673–677.

"Rising Seas Threaten Bay Marshes." Environment News Service, April 11, 2002. http://ens-news.com/ens/apr2002/2002L-04-11-09.html.

"Rising Tide: Who Needs Essex Anyway." *London Guardian*, June 12, 2003, 4.

Roberts, Callum M., Colin J. McClean, John E. N. Veron, Julie P. Hawkins, Gerald R. Allen, Don E. McAllister, et al. "Marine Biodiversity Hotspots and Conservation Priorities for Tropical Reefs." *Science* 295 (February 15, 2002): 1280–1284.

Roberts, Greg. "Great Barrier Grief as Warm-Water Bleaching Lingers." *Sydney Morning Herald*, January 20, 2003, 4.

Roberts, Paul. *The End of Oil: On the Edge of a Perilous New World.* Boston: Houghton-Mifflin, 2004.

Robock, Alan. "Pinatubo Eruption: The Climatic Aftermath." *Science* 295 (February 15, 2002): 1242–1244.

Robson, Seth. "Glaciers Melting." *Christchurch (New Zealand) Press*, February 26, 2003, 13.

"Rock Measurements Suggest Warming Is Global." Environment News Service, April 16, 2002. http://ens-news.com/ens/apr2002/2002L-04-16-09.html.

Rodbell, Donald T. "The Younger Dryas: Cold, Cold Everywhere?" *Science* 290 (October 13, 2000): 285–286.

Rodo, Xavier, and Francisco A. Comin, eds. *Global Climate: Current Research and Uncertainties in the Climate System.* Berlin: Springer-Verlag, 2003.

Roe, Nicholas. "Show Me a Home Where the Reindeer Roam." *London Times*, November 10, 2001. (Lexis).

Rogers, David J., and Sarah E. Randolph. "The Global Spread of Malaria in a Future, Warmer World." *Science* 289 (September 8, 2000): 1763–1766.

Rohling, Eelco J., Robert Marsh, Neil C. Wells, Mark Siddall, and Neil R. Edwards. "Similar Meltwater Contributions to Glacial Sea Level Changes from Antarctic and Northern Ice Sheets." *Nature* 430 (August 26, 2004): 965–968.

Rohter, Larry. "Antarctica, Warming, Looks Ever More Vulnerable." *New York Times*, January 25, 2005. www.nytimes.com/2005/01/25/science/earth/25ice.html.

———. "Deep in the Amazon Forest, Vast Questions about Global Climate Change." *New York Times*, November 4, 2003. www.nytimes.com/2003/11/04/science/earth/04AMAZ.html.

———. "Punta Arenas Journal: In an Upside-Down World, Sunshine Is Shunned." *New York Times*, December 27, 2002, A-4.

———. "A Record Amazon Drought, and Fear of Wider Ills." *New York Times*, December 11, 2005. www.nytimes.com/2005/12/11/international/americas/11amazon.html.

Root, Terry L., Jeff T. Price, Kimberly L, Hall, Stephen H. Schneider, Cynthia Rosenzweig, and J. Alan Pounds. "Fingerprints of Global Warming on Wild Animals and Plants." *Nature* 421 (January 2, 2003): 57–60.

Rosen, Yereth. "Alaska's Not-So-Permanent Frost." *Christian Science Monitor*, October 7, 2003, 1.

Rowan, Rob. "Thermal Adaptation in Reef Coral Symbionts." *Nature* 430 (August 12, 2004): 742.

Rowland, Sherwood, and Mario Molina. "Stratospheric Sink for Chlorofluoromethanes: Chlorine Atom-Catalyzed Destruction of Ozone." *Nature* 249 (June 28, 1974): 810–812.

Rowlands, Ian H. *The Politics of Global Atmospheric Change*. Manchester, U.K.: Manchester University Press, 1995.

Royer, Dana L., Scott L. Wing, David J. Beerling, David W. Jolley, Paul L. Koch, Leo J. Hickey, et al. "Paleobotanical Evidence for Near Present-Day Levels of Atmospheric CO_2 during Part of the Tertiary." *Science* 292 (June 22, 2001): 2310–2313.

Royse, David. "Scientists: Bush Global Warming Stance Invites Stronger Storms." Associated Press, October 25, 2004. (Lexis).

Rubbelke, Dirk T. G. *International Climate Policy to Combat Global Warming*. Cheltenham, U.K.: Edward Elgar, 2002.

Rubin, Daniel. "Venice Sinks as Adriatic Rises." Knight-Ridder News Service, July 1, 2003. (Lexis).

Rubin, Josh. "Toronto's Blooming Warm: Gardens, Golfers Spring to Life as Record High Nears." *Toronto Star*, December 5, 2001, B-2.

Ruddiman, William F. *Earth's Climate: Past and Future*. New York: Freeman, 2001.

Russell, Sabin. "Glaciers on Thin Ice: Expert Says Melting to Be Faster Than Expected." *San Francisco Chronicle*, February 17, 2002, A-4.

Russell-Jones, Robin. "Letter: Ozone in Peril." *London Independent*, December 7, 2000, 2.

Rutherford, Scott, and Steven D'Hondt. "Early Onset and Tropical Forcing of 100,000-Year Pleistocene Glacial Cycles." *Nature* 408 (November 2, 2000): 72–75.

Ryall, Julian. "Tokyo Plans City Coolers to Beat Heat." *London Times*, August 11, 2002, 20.

Ryan, Siobhain. "National Icons Feel the Heat." *Australia Courier Mail*, February 4, 2002, 1.

Ryskin, G. "Methane-Driven Oceanic Eruptions and Mass Extinctions." *Geology* 31 (2003): 737–740.

Sabadini, Roberto. "Ice Sheet Collapse and Sea-Level Change." *Science* 295 (March 29, 2002): 2376–2377.

Sabine, Christopher L., Richard A. Feely, Nicolas Gruber, Robert M. Key, Kitack Lee, John L. Bullister, et al. "The Oceanic Sink for Anthropogenic CO_2." *Science* 305 (July 16, 2004): 367–371.

Sadler, Richard, and Geoffrey Lean. "North Sea Faces Collapse of Its Ecosystem." *London Independent*, October 19, 2003, 12.

Sagan, Carl. *Billions and Billions: Thoughts on Life and Death at the Brink of the Millennium*. New York: Ballentine Books, 1997.

Sagarin, Raphael. "False Estimates of the Advance of Spring." *Nature* 414 (December 6, 2001): 600.

———, and Fiorenza Micheli. "Climate Change in Nontraditional Data Sets." *Science* 294 (October 26, 2001): 811.

Saleska, Scott R., Scott D. Miller, Daniel M. Matross, Michael L. Goulden, Steven C. Wofsy, Humberto R. da Rocha, et al. "Carbon in Amazon Forests: Unexpected Seasonal Fluxes and Disturbance-Induced Losses." *Science* 302 (November 28, 2003): 1554–1557.

Sample, Ian "Wouming Hits Tipping Point: Climate Change Alarm as Siberian Permafrost Melts for First Time Since Ice Age." *Manchester Guardian Weekly*, August 18, 2005, 1. www.guardian.co.uk/guardianweekly/story/0,12674, 1550685,00.html.

Sandalow, David B., and Ian A. Bowles. "Fundamentals of Treaty-Making on Climate Change." *Science* 292 (June 8, 2001): 1839–1840.

Santer, B. D., M. F. Wehner, T. M. L. Wigley, R. Sausen, G. A. Meehl, K. E. Taylor, et al. "Contributions of Anthropogenic and Natural Forcing to Recent Tropopause Height Changes." *Science* 301 (July 25, 2003): 479–483.

———, M. F. Wehner, T. M. L. Wigley, R. Sausen, G. A. Meehl, K. E. Taylor, et al. Response to Comment on "Contributions of Anthropogenic and Natural Forcing to Recent Tropopause Height Changes." *Science* 303 (March 19, 2004): 1771.

————, T. M. L. Wigley, G. A. Meehl, M. F. Wehner, C. Mears, M. Schabel, et al. "Influence of Satellite Data Uncertainties on the Detection of Externally Forced Climate Change." *Science* 300 (May 23, 2003): 1280–1284.

Sato, Makiko, James Hansen, Dorothy Koch, Andrew Lacis, Reto Ruedy, Oleg Dubovik, et al. "Global Atmospheric Black Carbon Inferred from AERONET." *Proceedings of the National Academy of Sciences* 100 (11) (May 27, 2003): 6319–6324.

Savill, Richard. "Tropical Fish Hooked on Channel Holidays." *London Daily Telegraph*, August 12, 2004, 7.

Scambos, T. A., J. A. Bohlander, C. A. Shuman, and P. Skvarca. "Glacier Acceleration and Thinning after Ice Shelf Collapse in the Larsen B Embayment, Antarctica." *Geophysical Research Letters* 31 (18) (September 22, 2004). doi:10.1029/2004GL020670.

Schar, Christoph, and Gerd Jendritzky. "Hot News from Summer 2003." *Nature* 432 (December 2, 2004): 559–561.

————, Pier Luigi Vidale, Daniel Luthi, Christoph Frei, Christian Haberli, Mark A. Linigier, et al. "The Role of Increasing Temperature Variability in European Summer Heatwaves." *Nature* 427 (January 22, 2004): 332–336.

Schiermeier, Quirin. "Gas Leak: Global Warming Isn't a New Phenomenon—Seabed Emissions of Methane Caused Temperatures to Soar in Our Geological Past, but No One Is Sure What Triggered the Release." *Nature* 423 (June 12, 2003): 681–682.

————. "Gulf Stream Probed for Early Warnings of System Failure." *Nature* 427 (February 26, 2004): 769.

————. "The Oresmen." *Nature* 421 (January 9, 2003): 109–110.

————. "Researchers Seek to Turn the Tide on Problem of Acid Seas." *Nature* 430 (August 19, 2004): 820.

————. "A Rising Tide: The Ice Covering Greenland Holds Enough Water to Raise the Oceans Seven Metres—and It's Starting to Melt." *Nature* 428 (March 11, 2004): 114–115.

Schimel, David, and David Baker. "Carbon Cycle: The Wildlife Factor." *Nature* 420 (November 7, 2002): 29–30.

————, J. I. House, K. A. Hibbard, P. Bousquet, P. Cials, P. Peylin, et al. "Recent Patterns and Mechanisms of Carbon Exchange by Terrestrial Ecosystems." *Nature* 414 (November 8, 2001): 169–172.

Schleifstein, Mark. "The Gulf [of Mexico] Will Rise, Report Predicts: Maybe 44 Inches, Scientists Say." *New Orleans Times-Picayune*, October 24, 2001, 1.

Schlesinger, W. H., and J. Lichter. "Limited Carbon Storage in Soil and Litter of Experimental Forest Plots under Increased Atmospheric CO_2." *Nature* 411 (May 24, 2001): 466–469.

Schmid, Randolph E. "Panel Sees Growing Threat in Melting Arctic." Associated Press, August 23, 2005. (Lexis).

———. "Warming Climate Reduces Yield for Rice, One of World's Most Important Crops." Associated Press, June 28, 2004. (Lexis).

Schmidt, Matthew W., Howard W. Spero, and David W. Lea. "Links between Salinity Variation in the Caibbean and North Atlantic Thermohaline Circulation." *Nature* 428 (March 11, 2004): 160–163.

Schmittner, Andreas, Masakazu Yoshimori, and Andrew J. Weaver. "Instability of Glacial Climate in a Model of the Ocean-Atmosphere-Cryosphere System." *Science* 295 (February 22, 2002): 1489–1493.

Schmitz, Birger. "Plankton Cooled a Greenhouse." *Nature* 407 (September 14, 2000): 143–144.

Schneider, Greg. "Taking No Chances: Disaster-Conscious Firms Treat Global Warming as a Reality." *Washington Post*, June 26, 2001, E-1.

Schneider, Stephen H. *Global Warming: Are We Entering the Greenhouse Century?* San Francisco: Sierra Club Books, 1989.

———. "No Therapy for the Earth: When Personal Denial Goes Global." In *Nature, Environment & Me: Explorations of Self in a Deteriorating World*, ed. Michael Aleksiuk and Thomas Nelson. Montreal: McGill-Queens University Press, forthcoming.

———. "What Is 'Dangerous' Climate Change?" *Nature* 411 (May 3, 2001): 17–19.

———, Armin Rosencranz, and John O. Niles, eds. *Climate Change Policy: A Survey*. Washington, D.C.: Island Press, 2002.

Scholes, R. J., and I. R. Nobles. "Storing Carbon on Land." *Science* 294 (November 2, 2001): 1012–1013.

Schrag, Daniel P., and Richard B. Alley. "Ancient Lessons for Our Future Climate." *Science* 306 (October 29, 2004): 821–822.

———, and P. F. Hoffman. "Geophysics, Life, Geology, and Snowball Earth." *Nature* 409 (January 18, 2001): 306.

———, and Braddock K. Linsley. "Corals, Chemistry, and Climate." *Science* 296 (April 12, 2002): 277–278.

Schrope, Mark. "Global Warming: A Change of Climate for Big Oil." *Nature* 411 (May 31, 2001): 516–518.

———. "Successes in Fight to Save Ozone Layer Could Close Holes by 2050." *Nature* 408 (December 7, 2000): 627.

Schultz, Martin G., Thomas Diehl, Guy P. Brasseur, and Werner Zittel. "Air Pollution and Climate-Forcing Impacts of a Global Hydrogen Economy." *Science* 302 (October 24, 2003): 624–627.

Schulze, Ernst-Detlef, Christian Wirth, and Martin Heimann. "Managing Forests after Kyoto." *Science* 289 (September 22, 2000): 2058–2059.

Schwartz, Stephen E., Thomas M. Smith, Thomas R. Karl, and Richard W. Reynolds. "Uncertainty in Climate Models." Letter to the Editor. *Science* 296 (June 21, 2002): 2139.

Schwartzman, David. *Life, Temperature, and the Earth: The Self-Organizing Biosphere.* New York: Columbia University Press, 2002.

"Scientists Feel Need for Urgency in Arctic Climate Research." Canadian Broadcasting Corporation, April 23, 2001. http://cbc.ca/cgibin/templates/view.cgi?/news/2001/04/23/nunavut_stu_01 0423.

Seager, R., D. S. Battisti, J. Yin, N. Gordon, N. Naik, A. C. Clement, et al. "Is the Gulf Stream Responsible for Europe's Mild Winters?" *Quarterly Journal of the Royal Meteorological Society* 128 (2002): 2563–2586.

"Sea of Japan 'Dead' in 350 Years." *Tokyo Daily Yomiuri*, November 30, 2000, 2.

Seelye, Katharine. "Environmental Groups Gain as Companies Vote on Issues." *New York Times*, May 29, 2003, C-1.

Seibel, Brad A., and Patrick J. Walsh. "Potential Impacts of CO_2 Injection on Deep-Sea Biota." *Science* 294 (October 12, 2001): 319–320.

Seidov, David E., Bernd J. Haupt, and Mark Maslin, eds. *The Oceans and Rapid Climate Change: Past, Present, and Future.* Washington, D.C.: American Geophysical Union, 2001.

Semenov, Vladimir A., and Lennart Bengtsson. *Modes of Wintertime Arctic Temperature Variability.* Hamburg, Germany: Max Planck Institut fur Meteorologie, 2003.

Semmens, Grady. "Ecologists See Disaster in Dwindling Water Supply." *Calgary Herald*, November 27, 2003, A-14.

Semple, Robert B. "A Film That Could Warm Up the Debate on Global Warming." Editorial Observer. *New York Times*, May 27, 2004. http://nytimes.com/2004/05/27/opinion/27THU3.html.

Serreze, M. C., J. A. Maslanik, T. A. Scambos, F. Fetterer, J. Stroeve, K. Knowles, et al. "A Record Minimum Arctic Sea Ice Extent and Area in 2002. *Geophysical Research Letters* 30 (2003). doi:10.1029/2002GL016406.

———, J. E. Walsh, F. C. Chapin, T. Osterkamp, M. Dyurgerov, V. Romanovsky, et al. "Observational Evidence of Recent Change in the Northern High-Latitude Environment." *Climatic Change* 46 (2000): 159–207.

Service, Robert F. "As the West Goes Dry." *Science* 303 (February 20, 2004): 1124–1127.

———. "The Carbon Conundrum." *Science* 305 (August 13, 2004): 962–963.

———. "Choosing a CO_2 Separation Technology." *Science* 305 (August 13, 2004): 963.

————. "The Hydrogen Backlash." *Science* 305 (August 13, 2004): 958–961.

"70 Cities in Indonesia Will Be Inundated." Antara, the Indonesian National News Agency, September 25, 2002. (Lexis).

Sevunts, Levon. "Prepare for More Freak Weather, Experts Say." *Montreal Gazette*, May 10, 2001, A-4.

Shackleton, Nicholas. "Climate Change across the Hemispheres." *Science* 291 (January 5, 2001): 58–59.

"Shanghai Mulls Building Dam to Ward Off Rising Sea Levels." Agence France Presse, February 9, 2004. (Lexis).

"Sharks in Alaskan Waters Could Herald Global Warming." Environment News Service, February 19, 2002. http://ens-news.com/ens/feb2002/2002L-02-19-09.html.

Sharp, David. "Study: New England's Winters Not What They Used to Be." Associated Press State and Regional News Feed, July 23, 2003. (Lexis).

Shaw, M. Rebecca, Erika S. Zavaleta, Nona R. Chiariello, Elsa E. Cleland, Harold A. Mooney, and Christopher B. Field. "Grassland Responses to Global Environmental Changes Suppressed by Elevated CO_2." *Science* 298 (December 6, 2002): 1987–1990.

Shepherd, Andrew, Duncan J. Wingham, Justin A. D. Mansley, and Hugh F. J. Corr. "Inland Thinning of Pine Island Glacier, West Antarctica." *Science* 291 (February 2, 2001): 862–864.

————, Duncan Wingham, Tony Payne, and Pedro Skvarca. "Larsen Ice Shelf Has Progressively Thinned." *Science* 302 (October 31, 2003): 856–859.

————, D. Wingham, and E. Rignot. "Warm Ocean Is Eroding West Antarctic Ice Sheet." *Geophysical Research Letters* 31 (23) (December 9, 2004). L23402, doi:10.1029/2004GL021106.

Sheppard, Charles R. C. "Predicted Recurrences of Mass Coral Mortality in the Indian Ocean." *Nature* 425 (September 18, 2003): 294–297.

Shindell, Drew T. "Climate Change: Whither Arctic Climate?" *Science* 299 (January 10, 2003): 215–216.

————, David Rind, and Patrick Lonergan. "Increased Polar Stratospheric Ozone Losses and Delayed Eventual Recovery Owing to Increasing Greenhouse-Gas Concentrations." *Nature* 392 (April 9, 1998): 589–592.

————, and G. A. Schmidt. "Southern Hemisphere Climate Response to Ozone Changes and Greenhouse Gas Increases." *Geophysical Research Letters* 31 (2004). doi:10.1029/2004GL020724.

————, Gavin A. Schmidt, Michael E. Mann, David Rind, and Anne Waple. "Solar Forcing of Regional Climate Change during the Maunder Minimum." *Science* 294 (December 7, 2001): 2149–2152.

———, G. A. Schmidt, R. L. Miller, and D. Rind. "Northern Hemisphere Winter Climate Response to Greenhouse Gas, Ozone, Solar, and Volcanic Forcing." *Journal of Geophysical Research* 106 (2001): 7193–7210.

Shwartz, Mark. "New Study Reveals a Major Cause of Global Warming—Ordinary Soot." *Stanford University Departmental News*, February 7, 2001. www.stanford.edu/dept/news/.

Siegenthaler, Urs, Thomas F. Stocker, Eric Monnin, Jakob Schwander, Bernhard Stauffer, Dominique Raynaud, Jean-Marc Barnola, Hubertus Fischer, Valrie Masson-Delmotte, and Jean Jouzel. "Stable Carbon Cycle Climate Relationship during the Late Pleistocene." *Science* 310 (November 25, 2005): 1313–1317.

Siegert, F., G. Ruecker, A. Hinrichs, and A. A. Hoffmann. "Increased Damage from Fires in Logged Forests during Droughts Caused by El Niño." *Nature* 414 (November 22, 2001): 437–440.

Siegert, Martin J., Brian Welch, David Morse, Andreas Vieli, Donald D. Blankenship, Ian Joughin, et al. "Ice Flow Direction Change in Interior West Antarctica." *Science* 305 (September 24, 2004): 1948–1951.

Sierra Club. "Global Warming: The High Costs of Inaction." 1999. www.sierraclub.org/global-warming/resources/innactio.htm.

Siggins, Lorna. "Warm-Water Anchovies Landed by Trawlers in Donegal Bay." *Irish Times*, December 12, 2001, 1.

Sigman, Daniel M., and Edward A. Boyle. "Glacial/Interglacial Variations in Atmospheric Carbon Dioxide." *Nature* 407 (October 19, 2000): 859–869.

Silver, Cheryl Simon, and Ruth S. DeFries. *One Earth, One Future: Our Changing Global Environment*. Washington, D.C.: National Academy Press, 1990.

Simons, Paul. "Weatherwatch." *London Guardian*, November 26, 2001, 14.

Singer, S. Fred, Donald Kennedy, and James D. Johnston.

Sirocko, Frank. "Paleoclimate: What Drove Past Teleconnections?" *Science* 301 (September 5, 2003): 1336–1337.

Siskind, David E., Stephen D. Eckermann, and Michael E. Summers, eds. *Atmospheric Science across the Stratopause*. Geophysical Monograph 123. Washington, D.C.: American Geophysical Union, 2000.

"Ski Resorts Get Creative to Battle Global Warming." Environment News Service, February 20, 2003. http://ens-news.com/ens/feb2003/2003-02-20-02.asp.

"Slowing Ocean Currents Could Freeze Europe." Environment News Service, February 21, 2002. http://ens-news.com/ens/feb2002/2002L-02-21-09.html.

Small, Jason. "Caskets, Bodies Surface during Frost Heaves in Yukon." *Whitehorse Star* in *Ottawa Citizen*, July 21, 2001, G-8.

Smith, Craig S. "One Hundred and Fifty Nations Start Groundwork for Global Warming Policies." *New York Times*, January 18, 2001, 7.

Smith, David C. "Marine Biology: Expansion of the Marine Archaea." *Science* 293 (July 6, 2001): 56–57.

Smith, Graeme. "Fishermen Fear the Worst over Cod: Parallels with Canada Collapse When 30,000 Jobs Were Lost." *Glasgow Herald*, October 23, 2002, 7.

Smith, L. C., G. M. MacDonald, A. A. Velichko, D. W. Beilman, O. K. Borisova, K. E. Frey, et al. "Siberian Peatlands a Net Carbon Sink and Global Methane Source since the Early Holocene." *Science* 303 (January 16, 2004): 353–356.

Smith, Lewis. "Falling Numbers Silence Cuckoo's Call of Spring." *London Times*, March 6, 2002. (Lexis).

Smith, Stanley D., Travis E. Huxman, Stephen F. Zitzer, Therese N. Charlet, David E. Housman, James S. Coleman, et al. "Elevated CO_2 Increases Productivity and Invasive Species Success in an Arid Ecosystem." *Nature* 408 (November 2, 2000): 79–82.

Smith, Steven J., Tom M. L. Wigley, and Jae Edmonds. "A New Route toward Limiting Climate Change?" *Science* 290 (November 10, 2000): 1109–1110.

Smith, Thomas M., Thomas R. Karl, and Richard W. Reynolds. "How Accurate Are Climate Simulations?" *Science* 296 (April 19, 2002): 483–484.

Smucker, Philip. "Global Warming Sends Troops of Baboons on the Run: Rising Temperatures and Humans Encroaching on Grasslands Are Endangering the Ethiopian Primates." *Christian Science Monitor,* June 15, 2001, 7.

Soden, Brian J., Richard T. Wetherald, Georgiy L. Stenchikov, and Alan Robock. "Global Cooling after the Eruption of Mount Pinatubo: A Test of Climate Feedback by Water Vapor." *Science* 296 (April 26, 2002): 727–730.

Solanki, S. K., I. G. Usoskin, B. Kromer, M. Schussler, and J. Beer. "Unusual Activity of the Sun during Recent Decades Compared to the Previous 11,000 Years." *Nature* 431 (October 28, 2004): 1084–1086.

"Solar Power Could Come from the Moon." Environment News Service, April 16, 2002. http://ens-news.com/ens/apr2002/2002L-04-16-09.html.

Sorensen, Eric. "The Letter We Can't See: Atmospheric Carbon Is One of the Worst Culprits in Global Warming." *Seattle Times*, April 22, 2001, A-1.

Souder, William. "Global Warming and a Toad Species' Decline." *Washington Post*, April 9, 2001. (Lexis).

Spahni, Renato, Jrme Chappellaz, Thomas F. Stocker, Laetitia Loulergue, Gregor Hausammann, Kenji Kawamura, Jacqueline Flückiger, Jakob Schwander, Dominique Raynaud, Valrie Masson-Delmotte, and Jean Jouzel. "Atmospheric Methane and Nitrous Oxide of the Late Pleistocene from Antarctic Ice Cores." *Science* 310 (November 25, 2005): 1317–1321.

Spalding, Mark. "Coral Grief: Rising Temperatures, Pollution, Tourism and Fishing Have All Helped to Kill Vast Stretches of Reef in the Indian Ocean.

Yet, with Simple Management, Says Mark Spalding, the Marine Life Can Recover." *London Guardian*, September 12, 2001, 8.

———. *World Atlas of Coral Reefs*. Berkeley: University of California Press, 2001.

Spears, Tom. "Antarctica Rides Global 'Heatwave': Continent's Warm Coast Causes Concern." *Ottawa Citizen*, August 8, 2001, A-1.

———. "Cold Spring Bucks the Trend." *Ottawa Citizen*, June 7, 2002, A-8.

Spero, Howard J., and David W. Lea. "The Cause of Carbon Isotope Minimum Events on Glacial Terminations." *Science* 296 (April 19, 2002): 522–525.

Speth, James Gustave. *Red Sky at Morning: America and the Crisis of the Global Environment*. New Haven, CT: Yale University Press, 2004.

Spotts, Peter N. "Trees No Savior for Global Warming." *Christian Science Monitor*, May 25, 2001.

Spray, Sharon L., and Karen L. McGlothlin. *Global Climate Change*. Lanham, MD: Rowman and Littlefield, 2002.

"Spray Cans Warming the Planet, One Dust-Busting Puff at a Time." *International Herald-Tribune* in *Tokyo Herald Asahi*, May 27, 2004. (Lexis).

Stainforth, D. A., T. Aina, G. Christensen, M. Collins, N. Faull, D. J. Frame, et al. "Uncertainty in Predictions of the Climate Response to Rising Levels of Greenhouse Gases." *Nature* 433 (January 27, 2005): 403–406.

Stark, Mike. "Assault by Bark Beetles Transforming Forests: Vast Swaths of West Are Red, Gray, and Dying. Drought, Fire Suppression, and Global Warming Are Blamed." *Billings Gazette* in *Los Angeles Times*, October 6, 2002, B-1.

Steig, Eric J. "Paleoclimate: No Two Latitudes Alike." *Science* 253 (September 14, 2001): 2015–2016.

Stein, Rob, and Shankar Vedantam. "Science: Notebook." *Washington Post*, November 12, 2001, A-9.

Stenni, Barbara, Valerie Masson-Delmotte, Sigfus Johnsen, Jean Jouzel, Antonio Longinelli, Eric Monnin, et al. "An Oceanic Cold Reversal during the Last Deglaciation." *Science* 253 (September 14, 2001): 2074–2077.

Stenseth, Nils Chr., Atle Mysterud, Geir Ottersen, James W. Hurrell, Kung-Sik Chan, and Mauricio Lima. "Ecological Effects of Climate Fluctuations." *Science* 297 (August 23, 2002): 1292–1296.

Stevens, William K. *The Change in the Weather: People, Weather, and the Science of Climate*. New York: Delacorte Press, 1999.

Stewart, Fiona. "Climate Change in the Back Garden." *The Scotsman*, November 20, 2002, 8.

Stewart, Richard B., and Jonathan B. Wiener. *Reconstructing Climate Policy beyond Kyoto*. Washington, D.C.: American Enterprise Institute, 2003.

Stiffler, Linda, and Robert McClure. "Effects Could Be Profound." *Seattle Post-Intelligencer*, November 13, 2003, A-8.

Stive, Marcel J. F. "How Important Is Global Warming for Coastal Erosion? An Editorial Comment." *Climatic Change* 64 (1/2) (May 2004): 27–39.

Stocker, Thomas F. "Climate Change: North-South Connections." *Science* 297 (September 13, 2002): 1814–1815.

———. "Global Change: South Dials North." *Nature* 424 (July 31, 2003): 496–497.

———, Reto Knutti, and Gian-Kasper Plattner. "The Future of the Thermohaline Circulation—A Perspective." In *The Oceans and Rapid Climate Change: Past, Present, and Future*, ed. Dan Seidov, Bernd J. Haupt, and Mark Maslin, 277–293. Washington, D.C.: American Geophysical Union, 2001.

Stoddard, Ed. "Global Warming Threatens 'Living Fossil' Fish: Coelacanths Have Existed 400 Million Years." Reuters in *Ottawa Citizen*, July 14, 2001, B-4.

Stokstad, Erik. "Defrosting the Carbon Freezer of the North." *Science* 304 (June 11, 2004): 1618–1620.

Stokstad, Erik. "Changes in Planktonic Food Web Hint at Major Disruptions in Atlantic." *Science* 305 (September 10, 2004): 1548–1549.

———. "Climate: River Flow Could Derail Crucial Ocean Current." *Science* 298 (December 13, 2002): 2110.

———. "Global Survey Documents Puzzling Decline of Amphibians." *Science* 306 (October 15, 2004): 391.

———. "States Sue over Global Warming." *Science* 305 (July 30, 2004): 590.

"Storm Warning." *Boston Globe*, May 23, 2004, N-17.

Stott, Lowell, Christopher Poulsen, Steve Lund, and Robert Thunell. "Super ENSO and Global Climate Oscillations at Millennial Time Scales." *Science* 297 (July 12, 2002): 222–226.

Stott, Peter A., and J. A. Kettleborough. "Origins and Estimates of Uncertainty in Predictions of Twenty-First Century Temperature Rise." *Nature* 416 (April 18, 2002): 723–726.

———, D. A. Stone, and M. R. Allen. "Human Contribution to the European Heatwave of 2003." *Nature* 432 (December 2, 2004): 610–613.

———, S. F. B. Tett, G. S. Jones, M. R. Allen, J. F. B. Mitchell, and G. J. Jenkins. "External Control of 20th Century Temperature by Natural and Anthropogenic Forcings." *Science* 290 (December 15, 2000): 2133–2137.

Streets, David G., Kejun Jiang, Xiulian Hu, Jonathan E. Sinton, Xiao-Quan Zhang, Deying Xu, et al. "Recent Reductions in China's Greenhouse Gas Emissions." *Science* 294 (November 30, 2001): 1835–1837.

Stroeve, J. C., M. C. Serreze, F. Fetterer, T. Arbetter, W. Meier, J. Maslanik, et al. "Tracking the Arctic's Shrinking Ice Cover: Another Extreme September Minimum in 2004." *Geophysical Research Letters* 32 (4) (February 25, 2005): L04501. http://dx.doi.org/10.1029/2004GL021810.

Struzik, Ed. "Fiery Future in Store for Forests If Climate Warms: Western Landscape Vulnerable: Study." *Edmonton Journal*, March 16, 2003, A-1.

Stuart, Simon N., Janice S. Cranson, Neil A. Cox, Bruce E. Young, Ana S. L. Rodrigues, Debra L. Fischman, et al. "Status and Trends of Amphibian Declines and Extinctions Worldwide." *Science* 306 (December 3, 2004): 1783–1786.

"Study of Ancient Air Bubbles Raises Concern about Today's Greenhouse Gases." *Omaha World-Herald*, November 25, 2005, 4-A.

"Study Reports Large-Scale Salinity Changes in Oceans: Saltier Tropical Oceans, Fresher Ocean Waters Near Poles Are Further Signs of Global Warming's Impacts on Planet." AScribe Newswire, December 17, 2003. (Lexis).

"Study Reveals Increased River Discharge to Arctic Ocean: Finding Could Mean Big Changes to Global Climate." AScribe Newswire, December 12, 2002. (Lexis).

"Study Shows Soil Warming May Stimulate Carbon Storage in Some Forests; Effect Would Slow Rate of Climate Change." AScribe Newswire, December 12, 2002. (Lexis).

Sturm, M., C. Racine, and K. Tape. "Climate Change: Increasing Shrub Abundance in the Arctic." *Nature* 411 (May 31, 2001): 546–548.

———. "Increasing Shrub Abundance in the Arctic." *Nature* 411 (May 31, 2001): 546–548.

Sturm, Matthew, Josh Schimel, Gary Michaelson, Jeffery M. Welker, Steven F. Oberbauer, Glen E. Liston, et al. "Winter Biological Processes Could Help Convert Arctic Tundra to Shrubland." *BioScience* 55 (1) (January 2005): 17–26.

"Suffocation Suspected for Greatest Mass Extinction." NewScientist.com News Service, September 9, 2003. www.newscientist.com/news/news.jsp?id=ns99994138.

"Summer Heat Wave in Europe Killed 35,000." United Press International, October 10, 2003. (Lexis).

Surendran, Aparna. "Fossil Fuel Cuts Would Reduce Early Deaths, Illness, Study Says." *Los Angeles Times*, August 17, 2001, A-20.

Sustainable World. "Global Warming 2." October 25, 2002. www.sustainableworld.org.uk/climate2.htm.

Svensen, Henrik, Sverre Planke, Anders Malthe-Sorenssen, Bjorn Jamtveit, Reidun Myklebust, Torfinn Rasmussen Eidem, et al. "Release of Methane from Volcanic Basin as a Mechanism for Initial Eocene Global Warming." *Nature* 429 (June 3, 2004): 542–545.

"Swedish Bogs Flooding Atmosphere with Methane: Thawing Subarctic Permafrost Increases Greenhouse Gas Emissions." American Geophysical Union

Website, February 10, 2004. www.scienceblog.com/community/article2366
.html.

Taalas, P., J. Damski, E. Kyro, M. Ginzburg, and G. Talamoni. "The Effect of
Stratospheric Ozone Variations on UV Radiation and on Tropospheric
Ozone at High Latitudes." *Journal of Geophysical Research* 102 (D1) (1997):
1533–1543.

———, E. Kyrö, K. Jokela, T. Koskela, J. Damski, M. Rummukainen, et al.
"Stratospheric Ozone Depletion and Its Impact on UV Radiation and on
Human Health." *Geophysica* 32 (1996): 127–165.

Tabazadeh, A., K. Drdla, M. R. Schoeberl, P. Hamill, and O. B. Toon. "Arctic
'Ozone Hole' in a Cold Volcanic Stratosphere." *Proceedings of the National
Academy of Sciences* 99 (5) (March 5, 2002): 2609–2612.

———, E. J. Jensen, O. B. Toon, K. Drdla, and M. R. Schoeberl. "Role of the
Stratospheric Polar Freezing Belt in Denitrification." *Science* 291 (March 30,
2001): 2591–2594.

Takahashi, Taro. "The Fate of Industrial Carbon Dioxide." *Science* 305 (July 16,
2004): 352–353.

———, Stewart C. Sutherland, Richard A. Feely, and Catherine E. Cosca.
"Decadal Variation of the Surface Water PCO_2 in the Western and Central
Equatorial Pacific." *Science* 302 (October 31, 2003): 852–856.

Tangley, Laura. "Greenhouse Effects: High CO_2 Levels May Give Fast-Growing
Trees an Edge." *Science* 292 (April 6, 2001): 36–37.

Tanner, Lawrence H., John F. Hubert, Brian P. Coffey, and Dennis P. McI-
nerney. "Stability of Atmospheric CO_2 Levels across the Triassic/Jurassic
Boundary." *Nature* 411 (June 7, 2001): 675–677.

Taubes, Gary. "Apocalypse Not." 1997. www.junkscience.com/news/taubes2
.html.

Taylor, James M. "Hollywood's Fake Take on Global Warming." *Boston Globe*,
June 1, 2004, A-11.

Teeman, Tim. "Trapped Explorer Must Build a Runway of Ice." *London Times*,
May 18, 2002. (Lexis).

Theobald, Steven. "Retailers Sweat It Out as Winter Sales Melt." *Toronto Star*,
December 6, 2001, D-1.

Thiemens, Mark H. "The Mass-Independent Ozone Isotope Effect." *Science* 293
(July 13, 2001): 226.

Thiessen, Mark. "Researchers Fighting 'Bad' Breath in Cattle." Associated Press,
June 8, 2003. (Lexis).

"Thin Polar Bears Called Sign of Global Warming." Environment News Service,
May 16, 2002. http://ens-news.com/ens/may2002/2002L-05-16-07.html.

Thomas, Chris D., Alison Cameron, Rhys E. Green, Michael Bakkenes, Linda J. Beaumont, Yvonne C. Collingham, et al. "Extinction Risk from Climate Change." *Nature* 427 (January 8, 2004): 145–148.

Thomas, R., E. Rignot, G. Casassa, P. Kanagaratnam, C. Acuòa, T. Akins, et al. "Accelerated Sea-Level Rise from West Antarctica." *Science* 306 (October 8, 2004): 255–258.

Thompson, Anne M., Jacquelyn C. Witte, Robert D. Hudson, Hua Guo, Jay R. Herman, and Masatomo Fujiwara. "Tropical Tropospheric Ozone and Biomass Burning." *Science* 291 (March 16, 2001): 2128–2132.

Thompson, Dan. "Experts Say California Wildfires Could Worsen with Global Warming." Associated Press, November 12, 2003. (Lexis).

Thompson, David W. J., and Susan Solomon. "Interpretation of Recent Southern Hemisphere Climate Change." *Science* 296 (May 3, 2002): 895–899.

———, and J. M. Wallace. "The Arctic Oscillation Signature in the Wintertime Geopotential Height and Temperature Fields." *Geophysical Research Letters* 25 (1998): 1297–1300.

———, and John M. Wallace. "Regional Climate Impacts of the Northern Hemisphere Annular Mode." *Science* 293 (July 6, 2001): 85–89.

Thompson, Lonnie G., Ellen Mosley-Thompson, Mary E. Davis, Keith A. Henderson, Henry H. Brecher, Victor S. Zagorodnov, et al. "Kilimanjaro Ice Core Records: Evidence of Holocene Climate Change in Tropical Africa." *Science* 298 (October 18, 2002): 589–593.

Thorpe, R. B., J. M. Gregory, T. C. Johns, R. A. Wood, and J. F. B. Mitchell. "Mechanisms Determining the Atlantic Thermohaline Circulation Response to Greenhouse Gas Forcing in a Non-Flux-Adjusted Coupled Climate Model." *Journal of Climate* 14 (July 15, 2001): 3102–3116.

Thurman, Judith. "In Fashion: Broad Stripes and Bright Stars: The Spring-Summer Men's Fashion Shows in Milan and Paris." *The New Yorker*, July 28, 2003, 78–82.

Tickell, Crispin. "Communicating Climate Change." *Science* 297 (August 2, 2002): 737.

"Time to Act on Nunavut Climate Change." Canadian Broadcasting Corporation, March 16, 2001. http://north.cbc.ca/cgi-bin/templates/view.cgi?/news/2001/03/16/16nunmoose.

Tolbert, Margaret A., and Owen B. Toon. "Solving the P[olar] S[tratospheric] C[loud] Mystery." *Science* 292 (April 6, 2001): 61–63.

Toman, Michael A., Ujjayant Chakravorty, and Shreekant Gupta. *India and Global Climate Change: Perspectives on Economics and Policy from a Developing Country.* Washington, D.C.: Resources for the Future, 2003.

Toner, Mike. "Arctic Ice Thins Dramatically, NASA Satellite Images Show." *Atlanta Journal and Constitution*, October 24, 2003, 1-A.

———. "Drought May Signal World Warming Trend." *Atlanta Journal and Constitution*, January 31, 2003, 4-A.

———. "Huge Ice Chunk Breaks Off Antarctica." *Atlanta Journal and Constitution*, March 20, 2002, A-1.

———. "Meltdown in Montana: Scientists Fear Park's Glaciers May Disappear within 30 Years." *Atlanta Journal and Constitution*, June 30, 2002, 4-A.

———. "Microscopic Ocean Life in Global Decline: Temperature Shifts a Cause or an Effect?" *Atlanta Journal and Constitution*, August 9, 2002, 3-A.

———. "Oceans' Acidity Worries Experts: Report: Carbon Dioxide on Rise, Marine Life at Risk." *Atlanta Journal and Constitution*, September 25, 2003. (Lexis).

———. "Temperatures Indicate More Global Warming." *Atlanta Journal and Constitution*, July 11, 2002, 12-A.

———. "Warming Rearranges Life in Wild." *Atlanta Journal and Constitution*, January 2, 2003, 1-A.

Toniazzo, T., J. M. Gregory, and P. Huybrechts. "Climatic Impact of a Greenland Deglaciation and Its Possible Irreversibility." *Journal of Climate* 17 (1) (January 1, 2004): 21–33.

Töpfer, Klaus. "Policy Forum. Climate Change: Whither after The Hague?" *Science* 291 (March 16, 2001): 2095–2096.

"Top U.S. Newspapers' Focus on Balance Led to Skewed Coverage of Global Warming, Analysis Reveals." AScribe Newswire, August 25, 2004. (Lexis).

Townsend, Mark. "Monsoon Britain: As Storms Bombard Europe, Experts Say That What We Still Call 'Freak' Weather Could Soon Be the Norm." *London Observer,* August 11, 2002, 15.

Travis, Davis J., Andrew M. Carleton, and Ryan G. Lauritsen. "Climatology: Contrails Reduce Daily Temperature Range." *Nature* 418 (August 8, 2002): 593–594.

Treaster, Jospeh B. "Gulf Coast Insurance Expected to Soar." *New York Times*, September 24, 2005. www.nytimes.com/2005/09/24/business/24insure.html.

Trenberth, Kevin E. "Climate Variability and Global Warming." *Science* 293 (July 6, 2001): 47.

———, ed. *Effects of Changing Climate on Weather and Human Activities.* Sausalito, CA: University Science, 2001.

———, Aiguo Dai, Roy M. Rassmussen, and David B. Parsons. "The Changing Character of Precipitation." *Bulletin of the American Meteorological Society* (September 2003): 1205–1217.

———, and Timothy J. Hoar. "The 1990–1995 El Niño-Southern Oscillation Event: Longest on Record." *Geophysical Research Letters* 23 (1) (January 1, 1996): 57–60.

———, Bruce A. Wielicki, Anthony D. Del Genio, Takmeng Wong, Junye Chen, Barbara E. Carlson, et al. "Changes in Tropical Clouds and Radiation." *Science* 296 (June 21, 2002): 2095.

Tromp, Tracey K., Run-Lie Shia, Mark Allen, John M. Eiler, and Y. L. Yung. "Potential Environmental Impact of a Hydrogen Economy on the Stratosphere." *Science* 300 (June 13, 2003): 1740–1742.

Tsuda, Atsushi, Shigenobu Takeda, Hiroaki Saito, Jun Nishioka, Yukihiro Nojiri, Isao Kudo, et al. "A Mesoscale Iron Enrichment in the Western Subarctic Pacific Induces a Large Centric Diatom Bloom." *Science* 300 (May 9, 2003): 958–961.

Tudhope, Alexander W., Colin P. Chilcott, Malcolm T. McCulloch, Edward R. Cook, John Chappell, Robert M. Ellam, et al. "Variability in the El Niño-Southern Oscillation through a Glacial-Interglacial Cycle." *Science* 291 (February 23, 2001): 1511–1517.

Turner, John, John C. King, Tom A. Lachlan-Cope, and Phil D. Jones. "Climate Change (Communication Arising): Recent Temperature Trends in the Antarctic." *Nature* 418 (July 18, 2002): 291–292.

Turner, John A. "Sustainable Hydrogen Production." *Science* 305 (August 13, 2004): 972–974.

"2001 the Second Warmest Year on Record." Environment News Service, December 18, 2001. http://ens news.com/ens/dec2001/2001L-12-18-01.html.

Tyndall, John. "On the Absorption and Radiation of Heat by Gases and Vapours, and on the Physical Connexion of Radiation, Absorption, and Conduction." *The London, Edinburgh, and Dublin Philosophical Magazine and Journal of Science*, 4th ser. (September 1861): 169–194.

Tyndall, J. "On Radiation through the Earth's Atmosphere." *The London, Edinburgh, and Dublin Philosophical Magazine and Journal of Science* 4 (1863): 200–207.

Uhlig Robert, "Mild Autumn Produces England's First 'Noble Rot.'" *London Daily Telegraph*, December 6, 2001, 1.

Unger, Mike. "Global Warming Hits Big Screen." *Annapolis (Maryland) Capital,* May 28, 2004, A-1.

United Nations Environmental Program. "Explorers in Antarctica Find Fresh Evidence of Global Warming." UNEP Press Release, January 19, 2001.

Unwin, Brian. "Rose in Sightings of Exotic Sea Life Enchant Devon and Cornwall: Fascinating Foreigners Drawn to the Cornish Riviera." *London Independent*, July 15, 2002, 7.

———. "Tropical Birds and Exotic Sea Creatures Warm to Britain's Welcoming Waters." *London Independent*, August 20, 2001, 7.

Urban, Frank E., Julia E. Cole, and Jonathan T. Overpeck. "Influence on Mean Climate Variability from a 155-Year Tropical Pacific Coral Record." *Nature* 407 (October 26, 2000): 989–1993.

Urquhart, Frank, and Jim Gilchrist. "Air Travel to Blame as Well." *The Scotsman*, October 8, 2002. (Lexis).

"U.S. Carbon Dioxide Emissions Decline." Environment News Service, July 2, 2002. http://ens-news.com/ens/jul2002/2002-07-02-09.asp.

"U.S. Carbon Dioxide Emissions Up Sharply." Associated Press in *Omaha World-Herald*, November 10, 2001, 7-A.

"U.S. Electricity Sector Makes Twice as Much Greenhouse Gas as Europe: Report." Agence France Presse, October 21, 2002. (Lexis).

"U.S. N.S.F.: Scientists Find Climate Change Is Major Factor in Drought's Growing Reach." M2 Presswire, January 12, 2005. (Lexis).

"U.S. Report Links Human Actions to Global Warming." Environment News Service, June 3, 2002. http://ens-news.com/ens/jun2002/2002-06-03-02.asp.

Utton, Tim. "The Rat Rampage." *London Daily Mail*, January 22, 2003. (Lexis).

"U.V. Radiation Linked to Deformed Amphibians." Environment News Service, June 21, 2002. http://ens-news.com/ens/jun2002/2002-06-21-09.asp#anchor4.

Vaughan, David G., J. Gareth, J. Marshall, William M. Connolley, John C. King, and Robert Mulvaney. "Climate Change. Devil in the Detail." *Science* 293 (September 7, 2001): 1777–1779.

———, G. J. Marshall, W. M. Connolley, C. L. Parkinson, R. Mulvaney, D. A. Hodgson, et al. "Recent Rapid Regional Climate Warming on the Antarctic Peninsula." *Climatic Change* 60 (3) (October 2003): 243–274.

Vavrus, Steve. "The Impact of Cloud Feedbacks on Arctic Climate under Greenhouse Forcing." *Journal of Climate* 17 (3) (February 1, 2004): 603–615.

Veizer, Jan, Yves Godderis, and Louis M. Francois. "Evidence for Decoupling of Atmospheric CO_2 and Global Climate during the Phanerozoic Eon." *Nature* 408 (December 7, 2000): 698–701.

Verburg, Piet, Robert E. Hecky, and Hedy Kling. "Ecological Consequences of a Century of Warming in Lake Tanganyika." *Science* 301 (July 25, 2003): 505–507.

Vergano, Dan. "Global Warming May Leave West in the Dust by 2050, Water Supplies Could Plummet 30 Per Cent, Climate Scientists Warn." *USA Today*, November 21, 2002, 9-D.

Verrengia, Joseph B. "In Alaska, an Ancestral Island Home Falls Victim to Global Warming: 'We Have No Room Left.'" Associated Press, September 9, 2002. (Lexis).

Verschuren, Dirk. "Global Change: The Heat on Lake Tanganyika." *Nature* 424 (August 14, 2003): 731–732.

Victor, David G. *Climate Change: Debating America's Policy Options.* New York: Council on Foreign Relations, 2004.

———. *The Collapse of the Kyoto Protocol and the Struggle to Slow Global Warming.* Princeton, NJ: Princeton University Press, 2001.

Vidal, John. "Better Get Used to It, Say Climate Experts." *London Guardian,* October 28, 2002, 3.

———. "The Darling Buds of February: Daffodils Flower and Frogs Spawn as Spring Gets Earlier and Earlier." *London Guardian,* February 23, 2002, 3.

———. "You Thought It Was Wet? Wait until the Asian Brown Cloud Hits Town: Extreme Weather Set to Worsen through Pollution and El Niño: Cloud with No Silver Lining." *London Guardian,* August 12, 2002, 3.

———, and Terry Macalister. "Kyoto Protests Disrupt Oil Trading." *London Guardian,* February 17, 2005, 4.

Vilhjalmsson, H. "Climatic Variations and Some Examples of Their Effects on the Marine Ecology of Icelandic and Greenland Waters, in Particular during the Present Century." *Rit Fiskideildar Journal of the Marine Research Institute* (Reykjavik, Iceland) 15 (1) (1997): 9–29.

Vinnikov, Konstantin Y., and Norman C. Grody. "Global Warming Trend of Mean Tropospheric Temperature Observed by Satellites." *Science* 302 (October 10, 2003): 269–272.

Visbeck, Martin. "The Ocean's Role in Atlantic Climate Variability." *Science* 297 (September 27, 2002): 2223–2224.

Vogel, Gretchen. "Central Europe Foods: Labs Spared as Climate Change Gets Top Billing." *Science* 297 (August 23, 2002): 1256.

Vogel, Nancy. "Less Snowfall Could Spell Big Problems for State." *Los Angeles Times,* June 11, 2001, A-1.

Voigt, Christiane, Jochen Schreiner, Andreas Kohlmann, Peter Zink, Konrad Mauersberger, Niels Larsen, et al. "Nitric Acid Trihydrate (NAT) in Polar Stratospheric Clouds." *Science* 290 (December 1, 2000): 1756–1758.

"Volcanic Eruptions Could Damage Ozone Layer." Environment News Service, March 5, 2002. http://ens-news.com/ens/mar2002/2002L-03-05-09.html.

Von Radowitz, John. "Antarctic Wildlife 'at Risk from Global Warming.' " Press Association News, September 9, 2002. (Lexis).

Wagner, Angie. "Debate over Causes Aside, Warm Climate's Effects Striking in the West." Associated Press, April 27, 2004. (Lexis).

Walker, Gabrielle. "Palaeoclimate: Frozen Time." *Nature* 429 (June 10, 2004). www.nature.com/cgi-taf/Dynapage.taf?file=/nature/journal/v429/n6992/full/429596a_fs.html.

"Walnuts and Vineyards." *London Times*, October 29, 2001. (Lexis).

Walsh, J. E., P. T. Doran, J. C. Priscu, W. B. Lyons, A. G. Fountain, D. M. McKnight, et al. "Climate Change (Communication Arising): Recent Temperature Trends in the Antarctic." *Nature* 418 (July 18, 2002): 292.

Walther, Gian-Reto. "Plants in a Warmer World." *Perspectives in Plant Ecology, Evolution, and Systematics* 6 (3) (2003): 169–185.

———. "Weakening of Climatic Constraints with Global Warming and Its Consequences for Evergreen Broad-Leaved Species." *Folia Geobotanica* 37 (2002): 129–139.

———, C. A. Burga, and P. J. Edwards, eds. *Fingerprints of Climate Change: Adapted Behaviour and Shifting Species Ranges*. New York: Kluwer Academic/Plenum, 2001.

———, Eric Post, Peter Convey, Annette Menzel, Camille Parmesan, Trevor J. C. Beebee, et al. "Ecological Responses to Recent Climate Change." *Nature* 416 (March 28, 2002): 389–395.

Wania, Frank, and Donald Mackay. "Tracking the Distribution of Persistent Organic Pollutants." *Environmental Science and Technology: News and Research Notes* 30 (9) (September 1996): 390–395.

"Warmer Climate Could Disrupt Water Supplies." Environment News Service, December 20, 2001. http://ens-news.com/ens/dec2001/2001L-12-20-09.html.

"Warming Could Submerge Three of India's Largest Cities: Scientist." Agence France Presse, December 6, 2003. (Lexis).

"Warming Doom for Great Barrier Reef." Australian Associated Press in *Hobart (Australia) Mercury*, February 16, 2002. (Lexis).

"Warming Streams Could Wipe Out Salmon, Trout." Environmental News Service, May 22, 2002. http://ens-news.com/ens/may2002/2002L-05-22-06.html.

"Warming Tropics Show Reduced Cloud Cover." Environment News Service, February 1, 2002. http://ens-news.com/ens/feb2002/2002L-02-01-09.html.

Warren, M. S., J. K. Hill, J. A. Thomas, J. Asher, R. Fox, B. Huntley, et al. "Rapid Responses of British Butterflies to Opposing Forces of Climate and Habitat Change." *Nature* 414 (November 1, 2001): 65–69.

"*Washington Monthly* Examines Effects of Global Warming on Bush's Ranch." *The Bulletin's Frontrunner*, August 23, 2002. (Lexis).

Watson, A. J., D. C. E. Bakker, A. J. Ridgwell, P. W. Boyd, and C. S. Law. "Effect of Iron Supply on Southern Ocean CO_2 Uptake and Implications for Glacial Atmospheric CO_2." *Nature* 407 (October 12, 2000): 730–733.

Watson, Jeremy. "Plan to Hold Back Tides of Venice Runs into Flood of Opposition from Greens." *Scotland on Sunday*, December 30, 2001, 18.

Watson, Robert T. "Climate Change: The Political Situation." *Science* 302 (December 12, 2003): 1925–1926.

Watt-Cloutier, Sheila. Speech to Conference of Parties to the United Nations Framework Convention on Climate Change. Milan, Italy, December 10, 2003. www.inuitcircumpolar.com/section.php?ID=8&Lang=En&Nav=Section.

Weart, Spencer R. *The Discovery of Global Warming*. Cambridge, MA: Harvard University Press, 2003.

Weaver, Andrew J., Oleg A. Saenko, Peter U. Clark, and Jerry X. Mitrovica. "Meltwater Pulse 1A from Antarctica as a Trigger of the Bolling-Allerod Warm Interval." *Science* 299 (March 14, 2003): 1709–1713.

———, and Francis W. Zwiers. "Uncertainty in Climate Change." *Nature* 407 (October 5, 2000): 571–572.

Webb, Jason. "Mosquito Invasion as Argentina Warms." Reuters, 1998. http://bonanza.lter.uaf.edu/~davev/nrm304/glbxnews.htm.

Weber, Bob. "Arctic Sea Ice Isn't Melting, Just Drifting Away: New Study." *Montreal Gazette*, April 25, 2001, A-8.

———. "Global Warming Up for Centuries: Snowcap Study." Canadian Press in *Toronto Sun*, November 28, 2002, 22.

Webster, Ben. "Boeing Admits Its New Aircraft Will Guzzle Fuel." *London Times*, June 19, 2001. (Lexis).

Webster, P. J., G. J. Holland, J. A. Curry, and H.-R. Chang. "Changes in Tropical Cyclone Number, Duration, and Intensity in a Warming Environment." *Science* 309 (September 16, 2005): 1844–1846.

Webster, Paul. "Russia Can Save Kyoto, If It Can Do the Math." *Science* 296 (June 21, 2002): 2129–2130.

Weinberg, Bill. "Hurricane Mitch, Indigenous Peoples and Mesoamerica's Climate Disaster." *Native Americas* 16 (3/4) (Fall/Winter 1999): 50–59.

Weiner, Jonathan. *The Next One Hundred Years: Shaping the Fate of Our Living Earth*. New York: Bantam Books, 1990.

Welch, Craig. "Global Warming Hitting Northwest Hard, Researchers Warn." *Seattle Times*, February 14, 2004. http://seattletimes.nwsource.com/html/localnews/2001857961_warming14m.html.

Weller, G. "Regional Impacts of Climate Change in the Arctic and Antarctic." *Annals of Glaciology* 27 (1998): 543–552.

Wesemael, Bas van, and Eric F. Lambin. "Carbon Sinks and Conserving Biodiversity." *Science* 294 (December 7, 2001): 2093.

Whalley, John, and Randall Wigle. "The International Incidence of Carbon Taxes." Paper presented at a Conference on Economic Policy Responses to Global Warming, Rome, Italy, September 1990.

Whipple, Dan. "Climate: The Arctic Goes Bush." United Press International, January 10, 2005. (Lexis).

White, James C. "Do I Hear a Million?" *Science* 304 (June 11, 2004): 1609–1610.

Whitfield, John. "Alaska's Climate: Too Hot to Handle." *Nature* 425 (September 25, 2003): 338–339.

Whittell, Giles. "Heat Wave: Russians Drink and Drown." *London Times*, July 23, 2001. (Lexis).

Whoriskey, Peter, and Joby Warrick. "Report Revises Katrina's Force: Hurricane Center Downgrades Storm to Category 3 Strength." *Washington Post*, December 22, 2005, A-3. www.washingtonpost.com/wp-dyn/content/article/2005/12/21/AR2005122101960.html.

"Why We're All Being Caught on the Hop by Global Warming." *Irish Independent*, July 17, 2004. (Lexis).

Wiechert, Uwe H. "Earth's Early Atmosphere." *Science* 298 (December 20, 2002): 2341–2342.

Wielicki, Bruce A., Takmeng Wong, Richard P. Allan, Anthony Slingo, Jeffrey T. Kiehl, Brian J. Soden, et al. "Evidence for Large Decadal Variability in the Tropical Mean Radiative Energy Budget." *Science* 295 (February 1, 2002): 841–844.

Wigley, T. M. L. "ENSO, Volcanoes, and Record Breaking Temperatures." *Geophysical Research Letters* 27 (2000): 4101–4104.

———, and S. C. B. Raper. "Interpretation of High Projections for Global-Mean Warming." *Science* 293 (July 20, 2001): 451–454.

"Wildfires Add Carbon to the Atmosphere." Environment News Service, December 9, 2002. http://ens-news.com/ens/dec2002/2002-12-09-09.asp.

Wilhite, Donald A. "Drought in the U.S. Great Plains." In *Handbook of Weather, Climate, and Water: Atmospheric Chemistry, Hydrology, and Societal Impacts*, ed. Thomas D. Potter and Bradley R. Colman, 743–758. Hoboken, NJ: Wiley Interscience, 2003.

Wilkin, Dwyane. "Global Warming Poses Big Threats to Canada's Arctic." *Nunatsiaq News*, November 21, 1997. www.nunatsiaq.com/archives/back-issues/71121.html#6.

———. "A Team of Glacial Ice Experts Say Mother Nature's Thermostat Has Kept the Eastern Arctic at about the Same Temperature since 1960." *Nunatsiaq News*, May 30, 1997. www.nunanet.com/~nunat/week/70530.html#7.

Williams, Brian. "Reef Down to Half Its Former Self." *Queensland (Australia) Courier Mail*, June 24, 2004, 11.

Selected Bibliography

Williams, Carol J. "Danes See a Breezy Solution: Denmark Has Become a Leader in Turning Offshore Windmills into Clean, Profitable Sources of Energy as Europe Races to Meet Emissions Goals." *Los Angeles Times*, June 25, 2001, A-1.

Williams, Frances. "Everest Hit by Effects of Global Warming." *London Financial Times*, June 6, 2002, 2.

Williams, Mary E., ed. *Is Global Warming a Threat?* San Diego: Greenhaven Press, 2003.

Williams, P. J. "Permafrost and Climate Change." *Philosophical Transactions of the Royal Society*, Series A, 352 (1995): 197–385.

Wilmsen, Steven. "Critters Enjoy a Baby Boom: Mild Winter's Downside Is Proliferation of Vermin." *Boston Globe*, March 30, 2002, B-1.

Wilson, Paul A., and Richard D. Norris. "Warm Tropical Ocean Surface and Global Anoxia during the Mid-Cretaceous Period." *Nature* 412 (July 26, 2001): 425–429.

Wilson, Scott. "Warming Shrinks Peruvian Glaciers: Retreat of Andean Snow Caps Threatens Future for Valleys." *Washington Post*, July 9, 2001, A-1.

"Wind Power Use Grows by 30 Per Cent." *London Guardian*, January 10, 2002, 15.

Wines, Michael. "Rising Star Lost in Russia's Latest Disaster." *New York Times*, September 24, 2002, A-11.

Winestock, Geoff. "How to Cut Emissions? E.U. Can't Decide." *Wall Street Journal*, July 13, 2001, A-9.

Wofsy, Steven C. "Where Has All the Carbon Gone?" *Science* 292 (June 22, 2001): 2261–2263.

Wohlforth, Charles. *The Whale and the Supercompter: On the Northern Front of Climate Change.* New York: North Point Press/Farrar, Straus and Giroux, 2004.

Wolf, Martin. "Hot Air about Global Warming." *London Financial Times*, November 29, 2000, 27.

Wolff, Eric W. "Whither Antarctic Sea Ice?" *Science* 302 (November 14, 2003): 1164.

———, Laurent Augustin, Carlo Barbante, Piers R. F. Barnes, Jean Marc Barnola, Matthias Bigler, et al. "Eight Glacial Cycles from an Antarctic Ice Core." *Nature* 429 (June 10, 2004): 623–627.

Woodard, Colin. "Wind Turbines Sprout from Europe to U.S." *Christian Science Monitor*, March 14, 2001, 7.

Woodcock, John. "Coral Reefs at Risk from Man-Made Cocktail of Poisons." *The Scotsman*, September 11, 2001, 4.

Woods, Audrey. "English Gardens Disappearing in Global Warmth: Will Be Replaced by Palm Trees." Associated Press in *Financial Post (Canada)*, November 20, 2002, S-10.

Woodward, F. I. "Carbon Cycle: Discriminating Plants." *Science* (March 30, 2001): 2562–2563.

Woodwell, George M. "Biotic Feedbacks from the Warming of the Earth." In *Biotic Feedbacks in the Global Climate System: Will the Warming Feed the Warming?* ed. George M. Woodwell and Fred T. MacKenzie, 3–21. New York: Oxford University Press, 1995.

———. "The Effects of Global Warming." In *Global Warming: The Greenpeace Report*, ed. Jeremy Leggett, 116–132. New York: Oxford University Press, 1990.

———. "Pregnancy in a Polluted World." *Science* 295 (February 1, 2002): 803.

———, and Fred T. MacKenzie, eds. *Biotic Feedbacks in the Global Climate System: Will the Warming Feed the Warming?* New York: Oxford University Press, 1995.

———, F. T. MacKenzie, R. A. Houghton, M. Apps, E. Gorham, and E. Davidson. "Biotic Feedbacks in the Warming of the Earth." *Climatic Change* 40 (1998): 495–518.

———, Fred T. MacKenzie, R. A. Houghton, Michael J. Apps, Eville Gorham, and Eric A. Davidson. "Will the Warming Speed the Warming?" In *Biotic Feedbacks in the Global Climate System: Will the Warming Feed the Warming?* ed. George M. Woodwell and Fred T. MacKenzie, 393–411. New York: Oxford University Press, 1995.

"World Wind-Power Capacity Marks Record Growth for 2002." Japan Economic Newswire, March 3, 2003. (Lexis).

Wright, J. D. "Climate Change: The Indonesian Valve." *Nature* 411 (May 10, 2001): 142–143.

"Yale University: Vast Majority of Americans Believe Global Warming Is 'Serious Problem.'" M2 Presswire, May 29, 2004. (Lexis).

Yam, Philip. "A Less Carbonated Earth." *Scientific American*, October 1999, 32.

Yohe, Gary, Natasha Andronova, and Michael Schlesinger. "To Hedge or Not against an Uncertain Climate Future?" *Science* 306 (October 15, 2004): 416–417.

Yoon, Carol Kaesuk. "Penguins in Trouble Worldwide." *New York Times*, June 26, 2001, F-1.

———. "Something Missing in Fragile Cloud Forest: The Clouds." *New York Times*, November 20, 2001, F-5.

Younge, Gary. "Bush U-turn on Climate Change Wins Few Friends." *London Guardian*, August 27, 2004, 18.

Yozwiak, Steve. "'Island' Sizzle, Growth May Make Valley an Increasingly Hot Spot." *The Arizona (Phoenix) Republic*, September 25, 1998. www.sepp.org/reality/arizrepub.html.

Selected Bibliography

Zachos, James C., Michael A. Arthur, Timothy J. Bralower, and Howard J. Spero. "Palaeoclimatology (Communication Arising): Tropical Temperatures in Greenhouse Episodes." *Nature* 419 (October 31, 2002): 897–898.

———, Nicholas J. Shackleton, Justin S. Revenaugh, Heiko Pälike, and Benjamin P. Flower. "Climate Response to Orbital Forcing across the Oligocene-Miocene Boundary." *Science* 292 (April 13, 2001): 274–278.

Zavaleta, Erika S., M. Rebecca Shaw, Nona R. Chiariello, Harold A. Mooney, and Christopher B. Field. "Additive Effects of Simulated Climate Changes, Elevated CO_2, and Nitrogen Deposition on Grassland Diversity." *Proceedings of the National Academy of Sciences* 100 (13) (June 24, 2003): 7650–7654.

Zeng, Ning. "Drought in the Sahel." *Science* 302 (November 7, 2003): 999–1000.

Zhang, Keqi, Bruce C. Douglas, and Stephen P. Leatherman. "Global Warming and Coastal Erosion." *Climatic Change* 64 (1/2) (May 2004): 41–58.

Zremski, Jeremy. "A Chilling Forecast on Global Warming." *Buffalo News*, August 8, 2002, A-1.

Zwally, H. Jay, Waleed Abdalati, Tom Herring, Kristine Larson, Jack Saba, and Konrad Steffen. "Surface Melt-Induced Acceleration of Greenland Ice-Sheet Flow." *Science* 297 (July 12, 2002): 218–222.

Zwiers, Francis W. "Climate Change: The 20-Year Forecast." *Nature* 416 (April 18, 2002): 690–691.

———, and Andrew J. Weaver. "The Causes of 20th Century Warming." *Science* 290 (December 15, 2000): 2081–2083.

INDEX

About the Author

BRUCE E. JOHANSEN is Professor of Communications and Native American Studies, University of Nebraska. He is the author of dozens of books, including *The Dirty Dozen: Toxic Chemicals and the Earth's Future, The Global Warming Desk Reference*, and *The Native Peoples of North America*. He is series editor of Praeger's *Native America: Yesterday and Today*.